INNOVATIVE APPROACHES TO TEACHING TECHNICAL COMMUNICATION

INNOVATIVE APPROACHES TO TEACHING TECHNICAL COMMUNICATION

Edited by

TRACY BRIDGEFORD
KARLA SAARI KITALONG
DICKIE SELFE

UTAH STATE
UNIVERSITY PRESS
Logan, Utah

Manufactured in the United States of America
Cover design by Barbara Yale-Read

Library of Congress Cataloging-in-Publication Data

Innovative approaches to teaching technical communication / edited by
Tracy Bridgeford, Karla Saari Kitalong, Dickie Selfe.
p. cm.
ISBN 0-87421-574-9 (pbck : alk. paper) – ISBN 0-87421-492-0 (ebook)
1. Communication of technical information–Study and teaching. I.
Bridgeford, Tracy, 1960- II. Kitalong, Karla Saari, 1952- III. Selfe,
Dickie, 1951-
T10.5.I518 2004
601'.4–dc22

2003021185

CONTENTS

INNOVATIVE APPROACHES TO TEACHING TECHNICAL COMMUNICATION

INTRODUCTION

Innovative Approaches to Teaching Technical Communication

Tracy Bridgeford
Karla Saari Kitalong
Dickie Selfe

The idea for this collection grew out of a discussion about humor in technical communication. Humor is usually proscribed in technical communication practice, both because it does not cross cultures well and because it may make complex and even dangerous technologies seem frivolous. When the three of us started paying attention to and collecting humor related to the technical communication field, we noticed that it is most often connected to Dilbert cartoons, the For Dummies genre of third-party software manuals, and Dave Barry–like rants about poorly written instructions. In short, the available humor about technical communication doesn't paint a pretty picture of our chosen profession.

Because we enjoy playful attitudes toward technical communication, we put aside the field's reservations about humor and began to ask ourselves how we might incorporate humor into our technical communication classrooms. Under what circumstances, we wondered, might humor be permissible or even desirable? How might it be used productively in the technical communication classroom?

In exploring that question, we concluded that perhaps what we were really talking about was how to demonstrate to technical communication students how *creative* the field could be. We were all teaching at Michigan Technological University at the time and had begun to notice that many students seemed disillusioned with the prospect of beginning their careers as *traditional* technical communicators. Writing instructional manuals for the computer industry or documenting ISO 9001 procedures for a government contractor seemed as dull as Dilbert's cubicle to them, especially when juxtaposed against the seemingly glamorous career prospects of Web or multimedia design. We didn't agree with them; in fact, our technical communication work as teachers, practitioners, and consultants has shown us its creative potential. For us and for the discipline at large, rendering complex information accessible and usable to an intended audience is interesting—even, at times, exhilarating. We wanted students to see that technical communication frequently involves creative effort, whether the project is a 500-page printed instruction manual or a digital media production. In other words, we were *not* willing to make the technocentric distinction many students seemed to be making between dull, document-based information and exciting, digital-media development.

Naturally, our first inclination was to blame ourselves. Had something about our previous teaching practices predisposed students to make unwarranted distinctions between project types? We thought we had carefully crafted our technical communication course assignments to show that the skills needed to design an interactive multimedia kiosk (a task students valued highly) were equally applicable in assembling a high-quality, well-indexed, and smoothly cross- referenced documentation set (a task many dismissed as boring). But somehow, our convictions were not getting through to them; moreover, another important idea that we wanted to emphasize—that success in all types of technical communication work necessitates strong and versatile writing skills—was apparently also lost on some undergraduate students. Imagine our chagrin when a graduating senior stated the opinion to one of our colleagues that writing is *retro!*

To impress our values on students and to convince them of the innovative practices available to technical communication professionals, the three of us developed a special topics course—Innovative Approaches to Technical Communication—which we offered to undergraduate scientific and technical communication (STC) majors at Michigan Technological University during the winter quarter 1997–1998. Our syllabus characterized the course as a way to engage more fully with the profession of technical communication by "considering a variety of attitudes, approaches, and practices." In the course design, we shifted students' attention (and our own) toward a more inclusive and creative view of the profession of technical communication. Course units advocated and modeled approaches that might not ordinarily be considered in technical communication. Students wrote, of course, but they also composed in other modalities such as mapping, drawing, scripting, acting, pantomiming, and MOOing. They read a variety of texts, including challenging theoretical texts not commonly assigned to undergraduates: Marilyn Cooper's (1996) *Technical Communication Quarterly* article on postmodern operator's manuals, James Porter and Patricia Sullivan's (1996) work on postmodern mapping, David Dobrin's (1989) "Armadillo Armor" article, and some articles on play theory (Huizinga 1990). To help students make the desired connections between their prior technical communication education and these new ideas, we assigned Janice Redish's (1988) "Reading to Learn to Do" article, Dorothy Nelkin's (1995) work on media representations of science and technology, and an excerpt from Robert Pirsig's (1974) *Zen and the Art of Motorcycle Maintenance.* To the stack of theoretical readings, we added newspaper editorials and articles about product assembly (Rooney 1997; Perelman 1976); cartoons from the Dilbert Zone (see www.dilbert.com); technology ads from popular magazines (Miller 1997) Will Weaver's short story titled "A Gravestone Made of Wheat" (1990); even some government documents—the Declaration of Independence and the Equal Rights Amendment.

The process of developing the reading list and the activities and techniques that went along with it ultimately reinforced our belief that the inspiration for our work as technical communication educators could come from almost anywhere, that our work could be as creative as we would allow it to be, and that we could approach the teaching and learning that goes on in technical communication courses in a variety of ways and from a variety of problem-solving perspectives. As we shared this discovery about teaching with students, they too came to realize that many theories, cultural artifacts, and issues could and should influence the work they do as technical communicators.

The learning that took place in the class does not represent a major paradigm shift in our students' thinking. But we are convinced that each student took away something profound. For example, one day, after reading a play theory essay and discussing the value of designing work spaces to facilitate playful interaction as well as serious labor, we each mapped our ideal work spaces. Michelle's included a desk for her best friend, collaborator, and longtime roommate, Jen, who had recently graduated and moved away. When Michelle shared the map with the rest of the class, she seemed a bit embarrassed by her dependence on her friend, and she hastily acknowledged the reality that she and Jen would never again work together. "Jen's desk" in a corner of Michelle's ideal work space stands as a metaphor for the lasting and palpable influence of a successful and pleasurable collaboration. Jen was not lost to Michelle: although she lived in another city and had chosen a technical communication career path that diverged from Michelle's, she would always be present in Michelle's work practices and habits of mind.

DEFINING TERMS AND ASSUMPTIONS

Program Names. Program naming conventions vary from institution to institution. Sometimes names signal the presence of deep and often contentious disciplinary boundaries. For instance, at some colleges, *communication* is a reserved title; its absence in English departments' course and program names may point to disciplinary turf wars. When we name our programs—technical and scientific communication, professional communication, professional or technical writing, business communication, and many others—we signal our institutional as well as our disciplinary allegiances. At base, though, the names are irrelevant. The editors of this collection are convinced that the techniques, approaches, technologies, and assignments suggested in this volume will transfer to programs across the disciplinary spectrum. Teachers in all these programs will find the materials described in this volume quite valuable.

That said, we assume that all teaching is highly contextualized and that the pedagogical explorations represented here will need to be adapted to the institutional, instructional, and material conditions that exist in each program and in each classroom. We hope that readers of this volume will not stop with the ideas and approaches presented herein, but will be inspired to continue to explore and test pedagogies from a variety of sources. One of the strengths of our diverse, but still related, disciplines is that we make it a point to attend carefully to the changing literacy practices that surround us, in academia and in society at large.

Critical. We use the term *critical* advisedly; the word and its seemingly commonsense derivations—critical reflection, critical pedagogy, critical thinking, critical approach—all carry a good deal of theoretical baggage and deserve to be, well, considered critically. However, we think that a critical approach to technical communication can yield quite practical results. In this volume we assume that *critical approaches* include any thoughtful, rhetorical, and culturally founded application of a technical communication process, practice, or approach to projects or documents (print and digital).

Innovative. Its prominence in the title of this collection makes *innovative* the term that most demands a common understanding. For us, to *innovate* means to introduce a new idea or to reintroduce an old idea, perhaps in a new way or in a new context. In this collection, then, an *innovative approach* is one that introduces, rearticulates, or creatively juxtaposes theories or practices, especially those not currently or commonly used within the context of technical communication teaching.

Some of the approaches described in this collection are unusual in technical communication pedagogy, curriculum development, or program design, even though they may be commonplace in other disciplines or contexts. Other approaches may be common in technical communication, but for some reason have not been published or otherwise officially entered into our professional discourse. The number of responses we received to our call for essays (almost three times as many as are included here) confirmed for us that innovative practices abound in technical communication classrooms, programs, and curricula.

To elicit the accounts of innovation we knew were "out there," we encouraged the authors to approach this collection as a way to share their most innovative instructional ideas, and we encourage readers to approach it as a catalyst for innovation in their work. We have asked authors to converse with their colleagues—the readers of this collection—by writing relatively short pieces that concisely but thoroughly describe their innovative activity or project in ways that allow readers to visualize implementation within a range of institutional contexts. We have also asked the authors to concentrate on the positive aspects of their approaches without sacrificing a critical discussion of problems posed by or reflected in their approaches, so that readers can more accurately project how approaches that appeal to them might fit into their institutional contexts.

THE THEORETICAL NEED FOR INNOVATION

From plain style to instrumental discourse to social constructivism and rhetoric, theoretically informed pedagogical discussions in technical communication have consistently focused on how best to prepare students for work (see, for example, Harris 1982, 630). This collection is timely and necessary as we consider our pedagogical responsibilities for preparing students for work now—at the beginning of the twenty-first century. Our thinking about pedagogy and the need for innovation (and hence, the selection of chapters for this volume) grows out of our reading of Etienne Wenger's (1998) *Communities of Practice;* Mary Sue Garay and Stephen Bernhardt's (1998) *Expanding Literacies: English Teaching and the New Workplace;* and James Paul Gee, Glynda Hull, and Colin Lankshear's (1996) *The New Work Order: Behind the Language of the New Capitalism.* From different perspectives, each of these books emphasizes the important aspects of what work means, as well as what it means to prepare students for work in the twenty-first century.

Although the new work order does not excuse students from developing an understanding of the forms of technical communication and how to execute them, Gee, Hull, and Lankshear(1996) point out that in this emerging work landscape, much more extensive and intensive literacies are expected than ever before. In essence—and this belief is something the technical communication discipline has known and accepted for a while—students' success depends on a commitment not only to classroom learning but also to lifelong learning. Workers in many fields can no longer consider their education to be complete upon graduating from college; today's college students need to develop learning strategies they can draw on throughout their careers, especially if they work in intensive, high-technology fields.

But, according to authors like Wenger, Garay and Bernhardt, and Gee, Hull, and Lankshear, a commitment to lifelong learning is just the tip of the proverbial iceberg. Students need, as well, to be able to respond quickly and effectively to continually changing local and global conditions and to rapid and unpredictable technological advancements. They need to develop and sustain a repertoire of learning and information management strategies and to reflect critically upon their choices and actions. In this environment, mastery of the forms and typical genres of technical communication is still necessary but is far from sufficient as a prerequisite for success in the new work order.

In short, as technical communication faculty, our charge to prepare students for work is complicated by the exigencies of the new work order. Our pedagogical practice is further complicated—and our job rendered perhaps even more important—in the wake of national and international events that swirl around us as this book goes to press. We are currently witnessing twin economic challenges—the volatility of high-tech industries and the upheaval of war and terrorism—that necessitate an attitude of innovation as an integral part of the technical communication teacher's toolkit for the classroom because it models the attitude of innovation that students need for *their* toolkits for the workplace.

We think you'll see that the *Innovative Approaches* authors—each in his or her way—focus on issues such as lifelong learning, the need to build workplace communities, changing workplace conditions, the globalization of technical communication work, and the increasing literacy demands being placed on technical communication practitioners.

ABOUT THIS COLLECTION: RESPONDING TO THE NEED FOR INNOVATION

Innovative Approaches to Teaching Technical Communication grows, then, out of students' apprehensiveness about their career paths, our efforts to understand the vagaries of the twenty-first century workplace, and our observation that many theoretical and disciplinary perspectives can *potentially* inform technical communication teaching, program administration, and curriculum development. Given pervasive changes in technology, the workplace, and cultural attitudes, new, dynamic, and flexible pedagogies seem warranted. *Innovative Approaches to Teaching Technical Communication* begins to address this need by demonstrating for technical communication faculty, graduate students, and program administrators the value of interrogating and innovating classroom and programmatic practices. The chapters were selected to highlight activities, projects, and approaches that have not been documented extensively in publications about technical communication teaching, curriculum development, or program administration. This book, therefore, offers the discipline another opportunity to energize its pedagogy and to critically examine current teaching practices.

The approaches described in this collection are practical, readily adaptable to a range of technological and institutional contexts, theoretically grounded, and pedagogically sound. They bring together a variety of scholars/teachers who expand an existing canon of publications about teaching technical communication (Fearing and Sparrow 1989; Selber 1997; Staples and Ornatowski 1997). Three objectives helped structure this collection. We wanted to

- capture a range of pedagogical perspectives that can inspire and invigorate technical communication teaching,

- present a variety of inventive, critical pedagogical practices for the technical communication classroom, and
- emphasize an array of partnership possibilities in technical communication pedagogy.

Using this framework, we looked for essays that demonstrated innovation in pedagogical perspectives, practices, and partnerships. We see this collection as broadening and making publicly accessible conversations already occurring in hallways, in faculty lounges, on listservs, and at conferences, but—for one reason or another—have not yet been made public. We hope that you will find as much inspiration in reading the pages of this book as we did in compiling them.

This collection is framed by the need for innovative pedagogical and curricular approaches that consider new perspectives, that describe new types of practices, and that exemplify new ways of establishing partnerships with industry. In each section, the authors think about and enact their pedagogical approaches by describing new ways of working and new strategies for adapting to changing workplace conditions.

Part One: Pedagogical Perspectives

The authors of these six chapters highlight and blend theories common to technical communication contexts with theories of technology, service learning, interdisciplinary and multicultural communication, and social interaction. Each chapter encourages reflective practices for both students and teachers. The approaches share the potential to foster in students the ability to think on their feet, to be flexible, and to respond to the needs of real audiences and clients.

In this section, six authors describe various perspectives for thinking about technical communication pedagogy. In the first chapter of this section, James Dubinsky examines the concept of service in a technical communication program and works toward a critical redefinition of that contested term. Sam Racine and Denise Dilworth describe an interactive television course in which they, together with their students, interrogated the television classrooms and technological systems that provided the context for their teaching and learning. Addressing the ever-increasing awareness of multicultural audience, Elaine Fredericksen showcases a bilingual professional writing program that capitalizes on students' fluency with both English and Spanish, thereby benefiting audiences and participants alike. W. J. Williamson and Philip Sweany explore the potential of discipline-specific service learning in technical communication as a general pedagogical theme and as an administrative innovation. Jeff Grabill writes about implementing and especially about sustaining a cross-curricular service-learning project in an urban university. Kathleen Yancey and her eleven colleagues from Clemson University describe the process of negotiation they went through in constructing their graduate reading list. Each of the chapters in this section energizes and enriches the paradigms to which we have become accustomed in technical communication.

Part Two: Pedagogical Practices

One of the chief concerns in technical communication has always been how best to teach it. In the classroom, theory and practice converge, each informing and shaping the other. Teachers often say that their classroom practices are affected by the theoretical landscapes they inhabit; in turn, their perspectives are borne out in the types of courses that they construct. Similarly, a particular theoretical innovation often grows out of effective, consistent classroom observations: teachers develop theories based on the types of activities they design, the classroom interactions they facilitate, and ultimately the students' reactions to their exploratory practices.

In this section of the collection, we draw together six chapters whose authors make a conscious effort to demonstrate the strategic interplay between theory and practice in their classrooms. They employ narrative, dramatic, genre, medical, cultural, and visual theories to inform their technical communication teaching. Each of the authors who contributed to this section has been careful to provide not only specific information about his or her particular theoretical perspective and how it plays out in the classroom but also to illustrate how the theories can be applied to other institutional contexts and adapted to individuals' own particular theoretical leanings. Tracy Bridgeford argues that using narrative ways of knowing enables students to more effectively understand and enact decision-making practices in technical communication contexts. James Kalmbach's chapter revisits the classic resume assignment in light of and in response to multiplying technological contexts. Barry Batorsky and Laura Renick-Butera describe their use of Brecht-inspired role-plays to foreground rhetorical problem-solving strategies. Karla Saari Kitalong suggests that the analysis of media representations can afford students an enhanced awareness of audience. Michael Zerbe employs cultural studies methodologies to help students gain a fuller and more critical understanding of medical communication. Dickie Selfe collaborates with his students to create technology-rich teaching environments informed by students' technology autobiographies.

Although some of these theoretical perspectives may sound familiar, these authors adopt unusual vantage points that promise to illuminate in new ways the study and teaching of technical communication.

Part Three: Pedagogical Partnerships

Academic technical communication programs are under constant pressure to develop and sustain mutually productive relationships with industry. There are many good reasons to establish such partnerships, not the least of which is to create a network of people who may hire students for internships and professional jobs. If, as conventional wisdom would have it, most job offers come as a result of a personal connection, then students' success in the job market may be enhanced by the corporate relationships that we cultivate. Moreover, as guest speakers, industry representatives can bring the atmosphere of the workplace into our classrooms, giving students—especially those with limited workplace experience—some insight into what is important in the so-called real world.

Industry partnerships can be helpful to financially strapped institutions, as well. Sponsored research projects, endowed chairs, equipment donations, and other financial or material collaborations often begin when individuals team up to work on projects, even if such projects are not directly focused on developing an industry-academia partnership. Finally, research in collaboration with corporate clients can lead to mutually beneficial experiences for companies and universities alike.

In Section 3, "Pedagogical Partnerships," seven authors describe a variety of frameworks for creating productive and long-lived partnerships with industry, describing and theorizing research in and around corporate settings to help readers make sense of this difficult and politically sensitive area of concern.

Stan Dicks and Brad Mehlenbacher write about engaging students in the usability testing of their campus's extension service Web pages. Craig Hansen suggests that service-based learning experiences—in conjunction with guided, critical reflections—are key to bridging practice and theory in technical communication. Christine Abbott explains a collaborative technical communication institute that brings together students, university faculty, and technical communication practitioners in an opportunity that blends theory and practice, study and networking. Annmarie Guzy's college-level technical communicators and Laura Sullivan's high school students come together in an online technical communication learning activity that has ramifications beyond the shared classroom experiences. Gary Bays reports on his interviews with corporate recruiters, during which he learned some strategies for initiating and sustaining workplace research projects that benefit both academia and industry. Billie J. Wahlstrom suggests that emerging venues for technical communication instruction demand that faculty develop new skills and new ways of envisioning instruction—what she calls "extreme pedagogies."

The idea of bringing together resonates throughout this section, as teachers, practitioners, corporate representatives, and students meet to share ideas and resources. The editors gratefully acknowledge the anonymous readers who reviewed this manuscript during its development period. Their careful and constructive comments allowed us to build on the strengths of the collection and minimize its weaknesses.

PART ONE

Pedagogical Perspectives

1

THE STATUS OF SERVICE IN LEARNING

James Dubinsky

During the last decade, a number of scholars/practitioners have explored the geographies of our fields, mapped the boundaries, and developed the landscape by building bridges (for example, Sullivan and Porter 1993; Blyler 1993; Forman 1993; and Allen 1992). One of the most important points in this discussion about identity has been the realization that to create a field of our own, we need to create our own major, one that will be independent and not subordinate. Sullivan and Porter (1993) explain that by

> conceiving of writing as a *major*, professional writing breaks with the dominant *service* identity assigned to composition. The development of professional writing as an academic entity signals a key conceptual shift: from the traditional notion of writing as ancillary to some other subject matter (i.e., writing as service to some other set of concerns—whether business, engineering, literature, or rhetoric/composition) to a recognition of writing as a discipline in its own right (i.e., a view that sees writing itself as a specialty area and as a subject of study). (405–6)

As they make a claim for professional writing's independence, Sullivan and Porter highlight *service* as one of the essential terms in the discussion. They link it to "the traditional notion of writing as ancillary to some other subject matter" and recognize that, for the most part, those of us who teach writing have been and continue to be marginalized (and to marginalize ourselves) because of connotations and history associated with *service*.

Yet, even as Sullivan and Porter (1993) long to break from that "dominant *service* identity" in order to get us to change our collective clothes, so to speak, and put on the mantle of respectability (which for them is associated with research), they recognize that what we do, at least to some extent, is indeed *service*. They explain that even with *writing* as a major, English departments "can continue their service functions and continue to be seen in that service role by some in the university" (406). Thus, despite their desire to cloak our "*service* identity," they do not dismiss it entirely. Service, deeply rooted in the spaces associated with writing, manages to maintain a presence in the landscape even as Sullivan and Porter work to re-map and re-present it.

In this chapter, my intention is not to argue with Sullivan and Porter's goal of achieving disciplinary status. I agree wholeheartedly that writing should be a discipline in its own right and a "subject of study." I disagree, however, that we need to break "with the dominant *service* identity" to accomplish those objectives. For that reason, I begin an inquiry into the concept of "service," a word many members of the profession of English language studies seem to want to keep hidden away like Rochester's wife in *Jane Eyre*. I examine some of the negative and positive connotations of the term when it is used as a modifier, such as those associated with being a "*service* discipline" and with the pedagogy of "*service*-learning," suggesting that we in the field of technical and scientific communication should bring *service* out of the attic, redefine it, and accept it as an integral component of our missions. In particular, I believe that service-learning, when used fully and reflectively, has the potential to enable us to move beyond negative modifiers. By accepting service as essential to what we do, we redraw the lines of the discussion, make the definitions we want to advance explicit, and take an active role not only in creating a curricular geography but also in assigning ourselves a place on the academic map that best represents us. Such an active role might enable us to achieve parity with other disciplines within the institutions of higher learning and avoid the fates of the non-European countries represented by European mapmakers, who were often marginalized, regardless of their actual size or status (Barton and Barton 1993).[1] More importantly, by accepting service as a key pedagogical goal, we revise our notion of scholarship and link practice and theory together in a manner reminiscent of classical Greece and Rome where rhetors worked to serve the public good.

THE FACES OF SERVICE

Use the term *service*, and you get many responses. On one hand, we have large, expansive definitions of service such as military service and service to country (the Kennedy inaugural speech or the 1993 National and Community Service Trust Act come to mind here), which are associated with volunteerism and duty. Linked to religious and social concepts, those who serve contribute to the public good and make their communities and country stronger (Bellah 1985; de Tocqueville 1974). On the other hand, there is a less expansive conception of service, the kind one expects while eating or shopping. Here, those who serve do so for pay or out of obligation or indenture, and there is little in the way of public advantage. The advantages are almost always private.[2]

In academe, the word *service* has a long history. Having just completed my annual faculty activity report (FAR), I'm well aware of the three criteria that others use to evaluate me: teaching, research, and service. And I know that at my school, a large, land-grant university, of the three criteria, service is the least valued. To use a common metaphor, academic work is seen as a stool with three legs. Unfortunately, in nearly every instance, service ends up being the shortest leg (Martin, 1977, vii; Mawby 1996, 49), and those who do more of it have less stable places to sit. The concept is accorded far less respect than its sister concepts of teaching and research (Boyer 1990). Many members of the academy see service as subordinate to teaching and research, so that even if they acknowledge that a primary mission of higher education is to serve, they argue that teaching and research, as the means to the end, should receive the most weight. To give an example, what should count is the research that leads to the discovery of a blight-resistant strain of corn or the teaching of how to plant and tend it. The planting and tending, the labor of bringing that plant to bear fruit, have far less weight.

In our discipline, the argument has long been that we don't have a subject of study. Our mission is not to discover new strains of corn or new processes for planting; our mission is to help those who do the discovering communicate their knowledge. Thus, most academics, including many of our colleagues in literature, justify their treatment of us because, for them, we exist in the less expansive mode. Our departments and courses exist because members of the university have a need for us. We are paid, so to speak, to provide others with services they need to do their work that will benefit the community. Returning to Sullivan and Porter's discussion, one can see that implicit in their desire to be rid of the term is the belief that when *service* is used as a modifier, what or who it modifies is second-rate (as in "Oh, they're a *service* discipline" or technical writing is "merely a *service* course"). Used in this manner, the term *service* falls into the second, less expansive mode; it is pejorative and condescending. Those involved in such work are more servants than equals, providing something necessary, yet something mechanical—a skill that other disciplines see as separate from their endeavors.[3]

SERVICE AS CONDUCT BECOMING A DISCIPLINE

In the military where I spent fifteen years of my adult life, I learned that there are actions or conduct that "become" one. These acts represent what is best about one's profession; they exemplify it, and members are expected to enact them by living in accordance with a code of conduct. So it goes for other professions as well, including that of teaching writing. We must know what is expected of us and live up to those expectations. Clearly, one of those expectations for teachers of professional writing is to teach students how to write well. Doing so is central to our profession; to deny otherwise is to bury our heads in the sand. More important, doing so—teaching students how to write well—is no easy task. To teach students how to write well, we need to understand what we're doing; we need to study both the act of writing and the teaching of the act of writing. We also need to study the effects those acts of writing have on others and use that knowledge to improve our teaching. Our work is a circle involving experiential learning—one that might be best expressed by Kolb's (1984) Learning Cycle, which combines concrete experience, through a reflective stage, on to an analytical stage, to a testing step, ending where it began, back at experience. This work, which I've argued elsewhere is like a Möbius loop, is essential to our field (see Dubinsky 1998). We must involve the act and art of teaching writing in the discussion. The strategy, however, is to argue that what we do, our labor, is inseparable from our teaching and our research. Thus, our service is of a piece with our scholarship.

The Service Mission of Higher Education

Rather than deny what has much truth (that we do, indeed, serve as Sullivan and Porter assert) or try to find a way to cloak or cover up that service with some "higher" calling such as study, we need to yoke the two concepts of service and study together. My first reason is that, as I've already stated, not all connotations of *service* are derogatory. Those that focus on "conduct tending to the welfare or advantage of another" *(OED)* are positive. These definitions seem in line not only with our field's historical role as the discipline responsible for literacy instruction but also with the mission of many institutions of higher learning, which is often associated with the concept of service.[4]

Relying on historical arguments and mission statements from colleges and universities, some scholars have been working to revive the concept of service. In *Scholarship Reconsidered,* Boyer (1990) argues for a redefinition of scholarship (the term associated with *research* that led to the uneven stool and a denigration of the concept of *service* when the modern university system was instituted).[5] He wants to see a broader definition of *scholarship,* one that encompasses what he calls the "scholarship of application" (16), a concept in which "service [is seen as] serious, demanding work requiring the rigor—and the accountability—traditionally associated with research" (22).

Along these same lines, a diverse group of educators has been working to create situations that require "reflection-in-action" (Schön 1983), involving a pedagogy that has come to be called service-learning, "an expanding . . . movement [that] educates students . . . for the benefit of society" (Henson and Sutliff 1998, 189). With this pedagogy, there is an emphasis on the scholarship associated with what Aristotle called productive knowledge (Miller 1984; Phelps 1991; Schön 1983), which links thought to action and theory to practice.

SERVICE-LEARNING: KEY TO REDEFINING SERVICE

These goals of redefining service and yoking the words *service* and *learning* speak directly to the issue presented by Sullivan and Porter (1993). How can we argue for independence and disciplinarity when one of the most difficult tasks we face as writing teachers is that we are not teaching a "subject of study" only? In nearly every course in nearly every technical communication curriculum I've examined, there is a practical component associated with the subject of study. We don't teach just document design; we teach how to design documents. We don't teach just about desktop publishing; we teach how to publish using tools available on our desktops. Even when we teach "theory" courses, all too often the theory revolves around the acts of writing (our own or those whom we teach or advise). As a result, there is a tension between how much emphasis we place on that practical component and how much we place on the subject of study.

The question at this point is how to make the argument about service and disciplinarity without giving up or relinquishing the connections, both historical and practical, to the work of teaching writing. One means is to consider the pedagogy of service-learning, which connects service to learning and unites practice and theory. Service-learning is a pedagogy in process and one that hasn't yet stabilized, having, according to one scholar, 147 different definitions (Kendall 1990). Despite the many definitions, there is quite a bit of agreement about the essential dynamics of the pedagogy, much of it codified at a national conference sponsored by the National Society for Internships and Experiential Education in 1991 (Giles, Honnet, and Migliore). The term refers to activities that combine work in the community with education. The "service" component is activity intended to assist individuals, families, organizations, or communities in need. The "learning" involves structured academic efforts to promote the development (intellectual and social) of the student. It also involves testing and reflection (thus, the link to the Kolb cycle presented previously). Although there is still much research to be done, there is statistical evidence that demonstrates an improvement in students' learning and commitment to a concept of citizenship (Markus, Howard, and King 1993; Cohen and Kinsey 1994; Parker-Gwin and Mabry 1998).

The pedagogy of service-learning elevates service's status to that of an equal with learning, one that doesn't have to be hidden away. It yokes two terms (learning and service) together that many have seen as oppositional; learning, the goal of higher education—knowledge for knowledge's sake—is literally tied by the hyphen to service. I argue elsewhere for the essential nature of the hyphen, but suffice it to say that the hyphen introduces an element of reciprocity, which results in a leveling of the legs of the stool (see Dubinsky 2002). The hyphen brings together *learning-by-doing* and *serving* (applying what one learns to one's community/society). One cannot have service-learning without some action, some activity conducted by the learners *for* and *with* other human beings.

Doing, however, is only part of the equation. There is an added dimension of ethical and social growth, fostered by reflection and conversation, designed to increase the students' investment in society. Consequently, the term service-learning implies both a type of program and a philosophy of learning (Anne Lewis, quoted in Kunin 1997, 155). What isn't readily apparent in the two words that compose the term is the key component of reflection, the glue that not only holds the two words together but also makes the whole far greater than the parts. Service-learning requires that students do more than just serve or learn; they must understand why and whom they serve and how that service fits into their learning (Bringle and Hatcher 1996; Sigmon 1994).

Service-learning, used fully and reflectively, helps students develop critical thinking skills; it also prepares students for the workplace in a more comprehensive way than many other pedagogical strategies because students apply what they've learned by working to develop reciprocal relationships with real audiences. These relationships, which are directed toward change not charity, enable students to meet their citizenship responsibilities (Dubinsky 2002). Service-learning pedagogy enables us to make our courses "a matter of *conduct* rather than of production" (Miller 1984, 23; Miller's emphasis). Students learn skills they'll use in the workplace, and they gain a practical wisdom *(phronesis)* that enables them to be critical citizens (Sullivan 1990).

TECHNICAL PROBLEM SOLVING OR SERVICE

To implement a pedagogy integral to creating an identity, one that creates relationships with people outside the academy and expands our classrooms beyond their traditional walls, we need to think about what we are doing and why. One of the key issues to resolve is whether we consider our work technical problem solving or service. Both have advantages, as outlined in a recent exchange between Johnson (1999) and Moore (1999), in which Moore is an advocate for "instrumental discourse," arguing that many technical communicators coming out of academic programs are held in low esteem because their communication *skills* are insufficiently developed, the same complaints made about the engineers in the early twentieth century (Kynell 1996). Moore's focus on instrumental skills, however, is the very focus that plays into the hands of those who want to belittle service. *If* what we do is defined by the job market only, if our work is measured by comma splices and the ability to use certain desktop publishing programs, then we are defining ourselves narrowly and not acknowledging the scope of what we do. If, however, we construct a definition of service so that we not only produce graduates who can use their skills for business and industry but also produce graduates who desire sincerely to use those skills to meet the community's needs and who have a desire to "share the common experience of learning about humans as they wrestle with technology in everyday situations" (Johnson 223), we are then embracing a version of service expansive and beneficial to society.

One key issue we need to consider concerns the attitudes we adopt and encourage students to adopt when we choose projects designed to help others. Although anxious to do good work, it is all too easy to adopt a charitable attitude that, while often well-intentioned, demonstrates that those doing the work feel superior because they have the answers to solve problems. Kahne and Westheimer (1996) describe this situation in terms of two competing models—charity and change—arguing that although both models may work, only the latter one enables people to work *with* others, to effect change and understand the underlying social issues and individual responsibilities. Linda Flower (1997) echoes this point. Drawing on John McKnight's (1995) analysis of social service policy, she explains that "community service has often rested on notions of philanthropy, charity, social service, and improvement that identify the community as a recipient, *client,* or patient, marked by economic, learning, or social deficits" (37; italics added).

For that reason, one of the key components of any service-learning project must be the underlying notion of the type of relationship that will exist between the class and the "client." Rather than encourage a "client" relationship, which is hierarchical, I encourage students to work *with* their organizations as partners. Although more complicated and requiring more of a commitment from the organization, students, and teacher, changing the relationship from a "client-consultant" to a three-way partnership changes dynamics that have a major effect on the outcomes in terms of the way students view problem solving and their roles as problem solvers and community members. Rather than going into a relationship with the assumption that the organization is the "problem" and the university will provide the answer, students understand the importance of working together with people to meet a need (McKnight 1995).

To illustrate, I describe how I learned to make the distinction between client and partner, between technical problem solving and service. In my earliest attempts to integrate service- learning, I emphasized to students the learning and the advantages that would accrue. The projects, following a model advocated by Huckin (1997), included an initial proposal, progress reports, the project itself (for example, Web sites, newsletters, annual reports), a reflection report, and an oral presentation to the class. We began the term working with seven clients, ranging from the New River Valley Free Clinic to the YMCA at Virginia Tech. By term's end, we had met most of the needs outlined by the clients, producing products that would be used. Students applied what they learned in class about issues such as audience analysis, design, and layout. They walked away with an item for their rèsumès and, in some cases, a product they could include in a portfolio. But there wasn't much the students could say in their reflection reports about service other than statements about how they felt good about "helping" and how much that "helping" would help them later in life. Nor could I say much. I didn't get to know the clients well. Although I spoke with all of them on the phone throughout the term, I never even met two of them face-to-face. In all but one case, neither the student teams nor I formed partnerships or learned much more about the organizations, the people who worked for them, and the people they served other than what we needed to know to complete the projects. The relationships were truly consultant-client relationships, with one exception, and that exception led me to reevaluate my pedagogy to focus more on the area between service and learning.

The team that opened my eyes took their project further than the others and helped me to understand the value of service. This team worked for Managing Information with Rural America (MIRA) in Christiansburg, a nearby town. MIRA,"a grantmaking initiative of the W.K. Kellogg Foundation's Food Systems/Rural Development program area, . . . seeks to draw upon the reservoirs of strength, tenacity, and civic commitment in rural communities and to help rural people use technology (electronic communications and information systems) as a tool to meet current and future challenges" (MIRA 2002); and the mission of this local chapter of MIRA was to make information accessible online. They asked for our help to create a newsletter; the team was asked to design it, write the first issue, and convert it to HTML.

Although their project was not different in kind or scope from the other projects, the advantage this team had was the energy of the larger team they joined. Most teams worked for organizations understaffed and desperate for help. In many cases, they did not have the expertise or the personnel to create the Web site or design the brochure. Thus, they asked for help and were glad to take it. They didn't have the time to supervise or, in some cases, even advise students. Thus, they were good candidates for client projects but not good ones for service-learning partnerships.

Although I didn't realize it at first, the team working with MIRA had a different situation. They became members of a larger project or team, a diverse group of local people interested in enhancing information exchange. They needed our student's expertise and help, but they wanted to work with these students, considering them as team members rather than consultants. In essence, they sought volunteers because of their expertise, and they expected that these volunteers would come to believe in the idea of the project.

Although the MIRA team had some internal problems and although working as part of a larger team had complications in terms of meeting deadlines, the students began to see that because they were involved in a dialogic, reciprocal relationship, they were learning more than just how to apply their technical skills. Because they framed their project in terms of the relationships they developed with the organization's members, they (and I) began to see a distinction between technical solutions and public action. Like the other teams, they did good work, but their approach and the assumptions they made about the organization and the people it served were different. They turned their work into service; no longer was it an act of experts providing solutions. Instead, they joined with others to solve problems that all of them could see. As one of the students put it, "My involvement with the service-learning project changed my outlook. My work with MIRA [Managing Information with Rural America] has had a profound impact on my commitment to volunteerism and has solidified my plans to become an active member of my community." For this student, service did not displace care; rather, service became a form of caring about the problem, the people, and the solution that he helped implement by "restructur[ing] the relationships of service around the Latin roots of the word—'feeling with' . . . [turning]

service from an act of charity or authority into an act of empathy that grasps an essential" (Flower 1997, 99).

GROWTH OF SERVICE-LEARNING IN PROFESSIONAL COMMUNICATION

When implemented in a manner similar to the one described previously, service-learning is an attractive pedagogy and a philosophy, one growing rapidly in all fields of education, at all levels. The term itself can be traced back to a group of pioneers in the late 1960s (Stanton, Giles, and Cruz 1999); and the "movement" based on the term began rather modestly in a variety of locations across the country by people with varied backgrounds. What brought them together were their beliefs that learning doesn't happen just in a classroom and education involves more than just knowledge for knowledge's sake. These individuals began grassroots organizations that have grown rapidly. Two—the Campus Outreach Opportunity League, a student-led advocacy group started in 1984, and Campus Compact, an organization of institutions of higher learning begun by the presidents of Brown, Georgetown, and Stanford Universities in 1985—have assisted the spread of service-learning, as has the creation of the Corporation for National and Community Service.[6]

Although service-learning traces its history back to the mid-1960s, until 1997, little had been written about service-learning and courses whose subject was communication. The first few articles centered on work done in composition and advanced composition courses (Crawford 1993; Mansfield 1993; Herzberg 1994). Then, Cushman (1996) expanded the concept by talking about how working in and for the community can help to mold rhetoricians who are agents for change. Her work was followed by other articles in journals in composition studies and business and technical communication that acknowledged not only the practical value of this pedagogy in terms of how it can improve students' ability to apply what they have learned but also its value toward increasing their sense of civic responsibility (Bush-Bacelis 1998; Dubinsky 2002; Haussamen 1997; Henson and Sutliff 1998; Huckin 1997; Matthews and Zimmerman 1999; Shutz and Gere 1998). These articles outlined methods for implementing this pedagogy, its problems, and its benefits. Although problems can range from students' failing to take service seriously and copping an attitude toward skills and the workplace to having difficulty learning to speak *for* the organizations they're working with, the benefits, when service-learning is implemented fully, are clear. Students, working with community partners to create Web sites, write promotional or informational materials (newsletters, fact sheets, brochures, annual reports), become more committed to their communities and believe that they are better prepared to write effectively.

This recent acknowledgment of service-learning in our field is linked, in part, to an understanding that we've been practicing forms of experiential learning for several decades. Huckin (1997), for instance, states that although he hadn't heard about service-learning prior to 1997, he knows of many colleagues who had been employing project-based learning as far back as the early 1980s. It is also linked to the idea that our field has roots in classical rhetoric and the work of rhetors in classical Greece and Rome involved a commitment to the polis, to society (Dubinsky 2002). Quintilian, for instance, talked about "ideal orators" willing to put their knowledge and skills to work for the common good, who revealed themselves "in the actual practice and experience of life" (quoted in Whitburn 1984, 228).

The growth of service-learning has led to a resurgence of the importance of service. In 1999, *Technical Communication Quarterly* devoted an entire issue of the journal to redefining the "service" course, talking in terms of radical new pedagogy (David and Kienzler 1999), multiple literacies (Nagelhout 1999), and situated learning (Artemeva, Logie, and St-Martin 1999). The combination of service-learning and a willingness to redefine the very course often used to illustrate our menial status points to a grassroots effort in our field to slow down or even halt the attempts to make a "conceptual shift" away from *service.* They also illustrate reasons why even those among us calling for a new identity do not choose to throw away the old clothes altogether.

REEXAMINING WHAT WE DO

The debate over the word *service* is old, tracing its roots in some ways to a separation between what is useful and higher knowledge, one highlighted by arguments such as Boyer's to redefine scholarship and Sullivan and Porter's to redefine the field of professional writing by breaking from a "*service* identity."[7] It is a debate whose time has come.

One point of focus (Ronald 1987) is to examine our relationships with those to whom we teach these "mechanical skills" in answer to one of the questions members of our discipline often ask when discussing the nature of our work: "[Are we] helping students get jobs and promotions or [are we] helping them become critical thinkers who can change and improve their professions?" The question seems directed at some of the issues embedded in the debate surrounding service. If all we do is help students get jobs, then perhaps all we do is provide something that, while necessary, is menial. If, on the other hand, we help students become better citizens, then perhaps there is substance to our discipline.

When such questions are asked, few consider the possibility that such a question might be a false dichotomy. Regardless of how you see the question, if we're "helping," then we're being of service. By "helping," we provide not only added value but also essential knowledge and skills that students, stakeholders, and society need. We provide a procedural knowledge that helps students get jobs and that gives them the tools to improve their workplaces and their organizational cultures.

By arguing for the value of "procedural knowledge," by making a case for pedagogies such as "service learning" and for the value of "the scholarship of application," we can begin to create an environment in which those who teach these courses don't have to feel like "factory workers." Yet, these arguments will be incomplete unless we explicitly address the tacit nature of the way others and we view "service."

David Russell (1991, 71) says, "Tacit traditions have remained tacit because academia had no shared vocabulary, no institutional forums for discussing discipline-specific writing instruction." Although he isn't talking about the concept of "service" per se, he is talking about the issue of literacy, which is a goal of our "service courses." The concept that teaching writing is a service to enable the other departments to focus on the "important" tasks of teaching their content areas is such a tacit tradition. It will remain tacit as long as the underlying debate about what *service* means is left unaddressed.

CONCLUSION

My immediate reason for addressing this topic is that I've been asked to develop a writing program to serve the needs of the many departments that believe they produce knowledge that benefits society. They see their primary mission as contained within my university's motto of "*Ut Prosim*" or "That I May Serve." They believe that the production of knowledge is separate from the rhetorical acts involved in such production. They see the service they do as essential and the service of those who teach writing as menial. In essence, they see a significant distinction between their kind of service and ours. What is worse, as evidenced by my brief poll, many of my departmental colleagues agree, perpetuating or extending what one scholar has called a "disciplinary Maginot Line" (Lanham 1983, 16).

To build programs in technical communication, achieve parity in institutions devoted to research, and circumvent the Maginot Line, the tacit tradition linked to the pejorative term of "service" needs to be brought out into the open for examination and discussion. We need to "see" that the forces that produced the universities and colleges many of us teach in are the same forces that created the need for our courses. We should wear the mantle of *service* proudly as we demonstrate the value of service to the university. We need not hide our relationship with service to claim disciplinarity. Instead, we should examine what James White (1985) calls "invisible discourse" (the implicit expectations that are part of a culture). To build and maintain programs ecologically balanced, one of our goals should be to make visible the expectations about service that our stakeholders and we hold.

I've begun to do just that by establishing a dialogue with members of other disciplines responsible for curriculum development. These discussions are playing a role in the redesign of our "service" courses, in which we're negotiating how we can integrate service-learning. By taking the lead and using service-learning pedagogy in my service courses and then publicizing the results through conversations, the Service-Learning Center, workshops, and university newsletters, I'm opening up a dialogue about the reasons for elevating service, which include "1) the civic, moral, and cognitive development of students, 2) the improvement of the quality of life of the community as a result of university work, and 3) the campus's contribution to democracy" (Bringle, Games, and Malloy 1999, 199).

Integrating service-learning into our curriculum and working to become reflective practitioners, while also encouraging students to reflect on the work they do and the situations contributing to that work, will add to our status. When we engage in service-learning, we engage in problem solving, and as Harry Boyte (1993) says, "Problem-solving . . . is not a narrowly utilitarian term" (63). We are offering a rhetorical education that has larger purposes. By asking students to go out into the community, we enable them to develop skills and insights by focusing on real problems of real people. They learn, by working in semester-long reciprocal relationships with organizations in the community, that our society isn't perfect and that there are many ways to effect change. Specifically, they learn that the skills we teach them, when applied with care, can cause things to happen, particularly when they see that most nonprofits depend on grants and funds that come through donations (often solicited via letters or newsletters) to keep them alive. They learn to become rhetoricians for change (Cushman 1996). Consequently, they learn the value of writing well, and they apply what we teach far more enthusiastically.

In every class I've taught in which I've integrated service-learning, even earlier ones when I failed to achieve a balance between service and learning, students have overwhelmingly found the projects valuable and asked if there would be other courses with that teaching and learning strategy.[8] For these students, the theoretical became practical because it was related to life. That said, the task of implementing this pedagogy isn't easy; finding the balance between service and learning is as difficult as finding the balance between theory and practice or workplace and academe. For teachers to bridge the gaps successfully, they must be aware of the need for balance between service and learning and of the potential problems associated with this need. In essence, they need to read teacher's stories such as those by Matthews and Zimmerman (1999) and Huckin (1997).

To maintain integrity and continuity of purpose, we also need to encourage our colleagues in disciplines such as communication, computer science, and graphic arts that become part of technical or professional communication programs to contribute to this dialogue about the social contexts for literacy and our obligations to students, stakeholders, and society. We should answer what service is, decide whom we serve and why, and determine what those answers mean to us and to those we serve. Once answered, we can define, develop, and defend the concept of service to argue effectively for our place in the academy. Doing so will enable us "to provide for an education for citizenship" (Newman 1985, 31), teach the process of deliberation and judgment essential to such an education (Sullivan 1990, 383), and empower students to effect change in their communities. If we accomplish those goals, we have served truly and expansively, and we'll have a unified vision of our discipline that is practical in the fullest sense and valued because of the ethical and political dimensions associated with it. Our colleagues in literature and other disciplines will see that the work we do extends beyond comma splices and forms; they'll see that our service not only teaches the skills they value but also enables students to function more fully in their workplaces.

2

BREAKING VIEWING HABITS

Using a Self-conscious, Participatory Approach in the ITV Classroom

Sam Racine
Denise Dilworth

As universities and colleges expand their reaches beyond their own walls and form partnerships with one another to bring greater variety and flexibility of courses to students within and between states, distance (or distributed) education gains prevalence. Distance education, by definition, relies on technological solutions to bring course content to students. The highly successful partnership between the University of Central Florida and Brevard Community College offers students courses taught in a variety of media, such as the World Wide Web, videotape, radio broadcasts, and interactive television. The Twin Cities campus of the University of Minnesota employs various technologies to deliver courses to and receive them from its partners—Southwest State University and the University Center Rochester—as well as its own distributed campuses in Duluth, Crookston, and Morris. In fact, a quick Internet search with the keywords "distance education" reveals the broadening scope of technology use in nontraditional learning spaces: Michigan, South Dakota, Kentucky, Idaho, and many other states and communities boast their ability to deliver classes to students at a distance.

Moreover, distributed teams in the corporate arena increasingly adopt emerging technologies to capitalize on the skills and knowledge of employees at remote locations while decreasing the need for those employees to travel—a concern of especial importance in the wake of recent terrorist attacks in New York and Washington, D.C.

As technical communicators, we appreciate the value of addressing the practical aspects of using technology in our teaching, but we also recognize the necessity of addressing the underlying social and political dynamics as well. If we do not address such concerns, we risk that students and faculty will approach "technology more as individual consumers than as collective producers" (Pew Higher Education Roundtable 1994, 3A). Focusing exclusively on the practical aspects of using a technology reinforces the primacy of the medium, rather than the educational and social needs of teachers and students. In fact, it encourages teachers and students to see the technology as inevitable, inescapable, and inflexible. As Johnson-Eilola (1997) notes, when we separate social concerns from technology, "users are discouraged from recognizing and understanding (let alone participating in) the ways technologies construct our lives" (98).

In this chapter, we propose a self-conscious, participatory approach to using technology that will allow teachers of technical communication to examine the power of a medium in collaboration with their students. In our discussion here, we focus on the technology of interactive television (ITV), a medium common in the technical communication classroom, distance education, and university-community college partnership programs, as well as corporate teleconferencing. Together, we can negotiate ITV as a cultural and historical artifact, whose realizations can be shaped according to the needs of the participants rather than the demands of the technology. Through a critical exploration, students and teachers can assume an agency denied to them through passive reception to the technology and can re-create the medium as they experience it.

Our approach includes the following three goals:

1. To understand ITV technology through collaborative discussion
2. To identify and examine underlying assumptions that define and limit our approaches to ITV
3. To explore and reinvent norms and conventions of ITV use in order to shape our own realizations of the technology

We recognize that using ITV is itself a form of technical communication, and we should therefore approach the medium not as a transaction but as an experience grounded in rhetorical sensitivity. Our approach will help students and teachers develop heightened media appreciation and, more importantly, it will encourage an informed, agency-assumed practice that can be applied beyond ITV to manage other tools still emerging in the workplace. Our intent is that even though we focus on

ITV, our discussion can help participants use other technologies more effectively as well.

THE TENSIONS OF SUBSTANTIVE AND INSTRUMENTAL VIEWS OF ITV TECHNOLOGY

As emerging technologies are increasingly employed in both academic and professional spheres, students need access to and education with them (Karis 1997; Shirk 1997; Tebeaux 1989; Zuboff 1988). To be successful, students, and we as instructors, need communicative and rhetorical skills, in addition to instrumental proficiency. ITV technology, as familiar and intuitive as it may first appear, is no exception to this rule.

We acknowledge the tension teachers often feel when the promise of a given technology turns into an encumbrance that counters our pedagogical convictions. As one of our instructors notes, "The technology is a barrier you're trying to mitigate, rather than a tool that will help you teach; the conditions of the ITV environment are simply not conducive to teaching." When the operational conditions of the technology and the values of our teaching clash, it precipitates a variety of responses from teachers: resignation to what we see as the constraints of the technology (Johnson-Eilola 1997); outright rejection of the technology (Gilchrist 1997); or adaptation of the technology to the teaching practices we value (Bruce and Rubin 1993). We must recognize, as Anson (1999) does, that the "key to sustaining our pedagogical advances in the teaching of writing, even as we are pulled by the magnetic forces of innovation, will be to take control of these technologies, using them in effective ways" (273). Our approach is intended to give teachers and students such control.

Unfortunately, the practical use of ITV often leads to discussions about technology as a constraint that limits our pedagogical approaches: microphones and cameras impede interaction, technology interferes with the creation and maintenance of collaborative groups, while sound delay disrupts interpersonal communication. What we find disconcerting about these discussions is that they are characteristic of what Johnson-Eilola (1997), drawing from the work of Andrew Feenberg, describes as *substantive* views of technology:

> In the substantive view, we have little choice about how to deploy specific technologies in specific instances: Once we have adopted technologies, they

> determine their own uses. . . . Like a highly communicable disease, technology remakes all it touches (and it touches all); the only alternative is to retreat. (102)

In other words, we see technology as part of a social and educational network in which we have little or no power, where the characteristics of the technology constrain our interactions and interfere with our teaching and learning. Yet, the instrumental view of technology provides even less of an alternative: in the instrumental view, "technology [is] a neutral tool for doing a person's bidding" (Johnson-Eilola 1997, 102). This view discounts the ways we shape—or can be shaped by—the technology itself and implies that all we need for success is better training, more refined skills, higher levels of competence.

A common outcome of both substantive and instrumental views is a disturbing lack of agency on the part of both students and instructors. Students and teachers feel trapped by the technology, unable to engage in natural learning and interaction. This feeling of powerlessness, however, is not inevitable. If we begin to recognize the ways in which we respond to the medium and shape ourselves to its features and capabilities, rather than vice versa, we can begin to shape technology to our needs and create the medium as we use it.

The approach we advocate is one that encourages instructors and students to identify and reflect upon influences shaping our uses of technology. It forces us out of passive, substantive ways of thinking about and dealing with the technology into active, agency-driven roles that will help us shape the way we and students use technology in the classroom. As Wahlstrom (1997) notes, "Without a sense of agency, [students and, we add, instructors] become technological determinists, failing to identify opportunities when they could initiate change" (131). Bruce and Rubin (1993) concur, adding that social, cultural, and economic realities "manifest themselves in details of classroom organization, availability of resources, mandated curricula, teacher preparation, . . . and so on. These factors shape the possibilities for change in the classroom" and should, we argue, be an important part of classroom discussion and activity (Bruce and Rubin 1993, 5). In the next section, we discuss our approach more fully by demonstrating how our three goals are met and include activities and assignments that can be adapted to multiple ITV configurations as well as to other media.

SELF-CONSCIOUS, PARTICIPATORY APPROACH IN ACTION

We want to be clear that we deny neither the necessity nor the value of instrumental competency. In fact, we begin our approach by advocating that teachers and students learn how the technology operates, including its configuration and the physical limitations of its use. We see our approach as one that, ultimately, is practical, in the fullest, most rhetorical sense of the word. Miller (1989) suggests that we understand "practical rhetoric as a matter of *conduct* rather than of production, as a matter of arguing in a prudent way toward the good of the community." Such a view of practical, or practice-oriented, rhetoric as conduct will allow technical communication teachers to "promote both competence and critical awareness of the implications of competence" (23). Competence thus becomes one layer of a complex context. Understanding the complexities of that context through self-conscious deliberation and active participation—in the medium as well as in the classroom activities—encourages students not only to obtain and maintain skills but also to understand what having those skills means for them as members of our profession.

Understand ITV Technology through Collaborative Discussion

Before we can exploit any medium, we need to understand how it works. We find that at the heart of students' concerns about interacting in the new class environment is fear of the unknown. Never has a student new to ITV entered the room and moved boldly to the front row without pausing and seeking reassurance. Faced with rows of monitors, microphones, and a glass-walled technicians' booth, most students back, wide-eyed, out of the room and recheck the room number before returning to slink cautiously into a back row seat. The first step in our approach to teaching ITV, then, is to remove this fear by exploring the technology we see. We have found that many difficulties in using ITV can be avoided if all participants begin with a clear understanding of what ITV is and how it functions, as well as its capabilities and possible drawbacks. For example, in ITV classrooms, people who can't be heard often ask that the microphones be "turned up," as if they work as amplifiers. In truth, sound levels do not involve volume control, and adjustments are best made by moving the position of the microphone relative to the speaker. To create an effective orientation—both to the technology and its potential for use—we suggest teachers incorporate the following activities: group discovery to focus on the characteristics of the medium; demonstration of how the technology functions; and the modeling of our experiences of self-discovery to demonstrate an effective model for exploring technology.

Students need to understand right away that the space they are in is not a traditional classroom and that the medium is not commercial television. To this end, we advocate using an activity that focuses on the characteristics of the medium, developed by one of our instructors who teaches oral communication via ITV. At the beginning of the first class meeting, this instructor asks students to write their answers to one or more compelling questions (for example, "If you could work in any situation possible, what would it be and why?") and share them with the class. After students complete this exercise, she informs them that they have just completed their first ITV presentation and then asks that together they generate a list of "the unique characteristics of the ITV medium." Without fail, they are able to identify a full set of attributes that often characterize ITV: delays in delivery, voices canceling each other out, perceptions that people are not real, and so forth. We find that the students' experience of discovering ITV characteristics on their own proves more effective than just being told about those characteristics. Students are engaged, they speak to each other across sites, and because their perceptions are acknowledged and validated, they become generally more confident and less intimidated.

We suggest that instructors further take the mystery out of the classroom technology by demonstrating the controls and showing "the man behind the curtain" in more than a metaphorical sense. Students and instructors can decrease their anxiety by understanding how technology creates the characteristics they perceive, and an effective demonstration and explanation of controls can do much to remove the fear of the unknown or unexpected. A guided exploration of the control booth allows students to become more fully aware of what they usually just sense is happening around them and helps students to understand what to expect during class transmission. For example, a demonstration can explain how and why cameras move when students speak, the change in monitor display resulting from voice-activation, and the switch from overhead to straight-on cameras. Often, the technician is the person who can best explain and demonstrate how the technology works, but unfortunately, ITV participants have a tendency to ignore the technician as a contributing member of the community within the classroom. Encouraging a dialogue with technicians can help students feel more confident in front of the cameras and more comfortable in making requests regarding how they see themselves and others during ITV transmission. We encourage teachers and students to talk with technicians throughout the class and recognize that, whether through the technician or their own control, they can make changes as to how the technology is used.

We also believe that one of the best ways to support student learning is to model our experiences of self-discovery and exploration. Often, teachers who work flawlessly with ITV and produce polished presentations intimidate students with their expertise. Therefore, we remind teachers to share their stories of learning with students. To best learn about ITV, instructors should experience it both as students and as teachers and, while participating in these roles, record their observations about perceptions and interactions so they can share their discoveries with students. We suggest attending ITV workshops, touring the ITV classroom and experimenting with the controls, attending meetings of other courses, and, in general, gleaning as much information as possible by engaging in the role of student as well as teacher. We tell students about our initial perceptions of the technology and then describe how these evolved as we came to witness how and why our perceptions were rejected or confirmed. This modeling helps students to consider the dominant narratives they bring to technology.

By positioning ourselves as active and self-conscious explorers of the medium, we hope to encourage the same exploration and reflection in students. Our intent is that the purpose of orientation is not only to describe how the technology works but also to show that the medium is there for us to challenge and exploit, not to shy away from or fear. Talking with technicians and modeling our interest in learning about the technology help students grasp Miller's concept of seeing practical rhetoric as conduct, rather than as passive acceptance and application of rules.

Identify and Examine Underlying Assumptions Defining ITV

As we stated earlier, participants need to understand that ITV is not a traditional classroom or commercial TV. Similarities between these "old" media and the "new" medium of ITV serve only to confuse us—and to invite comparisons between the two that inhibit our understanding and perpetuate unproductive, substantive responses to ITV. Students must not only recognize that they respond to ITV in conventionalized ways but must also understand as well why they respond in those ways.

Our second goal for our approach is to persuade teachers to take time to explore students' assumptions about ITV technology as well as their own. All participants need to understand the experience of the whole class community, and to this end, teachers need to promote activities that enrich students' understanding of what it means to be a productive, successful professional. We have found that with a self-conscious attention to student-centered pedagogy, assumptions about constraints can be turned easily into opportunities for professional enhancement in terms of both conduct and skill building.

One of students' most common assumptions about ITV is that it is little different from commercial television, or as one student puts it: "When I watch TV, I zone out." Indeed, the "interactive" aspect of ITV is the one least intuited by students. The screens in front of the class are not televisions in the common understanding of the term, but instead are monitors, even though they are the size and shape of screens commonly found in dorms and living rooms. Perceived as commercial television, students are unwilling to interact with the people they see; after all, "talking back to the television" is not acceptable social behavior. Further, feelings of "watching television" reinforce the kind of passive behavior that is the antithesis of the active exploration we advocate. Interaction will not occur "naturally" until such assumptions are exposed and reshaped.

Another response to the medium, often a result of orientation materials that stress the avoidance of noise, is that students feel "like we're in church." They feel they must sit quietly and not fidget or make extraneous noise, or else the camera will zoom in on them and they will be placed in a very negative "spotlight." Such perceptions inhibit community building, the free exchange of ideas, and dialogues with others. Again, we've found that when students have the opportunity to discuss these fears and have the opportunity to see the reality as far less dramatic than their assumptions, they are generally more relaxed, engage the course content, and contribute more freely.

One of the best ways to understand the experience of the whole class community is to share each other's physical context. In the traditional classroom, we not only see our surroundings but also feel relatively certain that what we see and experience is pretty close to what students see and experience. This assumption is one we can't afford to make in the ITV classroom: many different classroom experiences exist, even within the confines of a single class. Each room has different monitor and microphone configurations, and the experience at our locale might be vastly different from that of our distant students, leading to unnecessary misunderstandings and unproductive misperceptions. For instance, we've all heard stories from teachers about how contrary students from distant sites can be: students move away from cameras so we can't see them or what they're doing. Our distant students, however, tell a much different story of what's happening: "We get tired of seeing ourselves on the monitor going out, so we move back from the camera to get a break." At the distance sites with which we have contact, there are two monitors at the front of the room: one that shows whoever is speaking at other sites and another that shows all the local students in their seats—a relentless (and distracting) mirror of their own activities.

Our distant students tell us of other instructor assumptions as well. Once we have gone through the process of sound and video checks with the local technician, we often take off at a run to make sure we have enough time to complete lessons and activities before the class ends and the monitors go blank. We often believe that, once checked, sound and transmission will remain fixed, that distant sites will continue to hear us throughout the class without our ever checking to ensure that they can. After all, in traditional classrooms, continued attention to sound, once established, is unnecessary. As our distant students tell us, however, this stability is not the case with ITV: "Instructors and the students at other sites don't realize we can't hear what they're saying, and we don't want to be the ones to interrupt. They need to check to see if they're being heard."

Taking the time simply to talk with students about the conditions created by the technology can be enlightening and can lead to problem-solving discussions that help students create their own solutions to the issues they identify and to which they respond. It can also open up avenues for collaboration with the technicians and for distant students' establishing themselves as site experts. We suggest that, early in the course, students at each site give a virtual "tour" of their locale, introducing the local technician and narrating the layout and the experience of working within that environment. This activity increases students' sensitivity to the multiple physical and rhetorical situations in which their class community must function and respond. It also prepares them for the variety of technological contexts they are likely to encounter beyond the classroom in corporate settings.

Explore and Reinvent Norms of ITV Use to Shape Our Realizations of the Technology

Our third goal for self-conscious, participatory approaches to ITV recalls Bruce and Rubin's (1993) distinctions between a technology's idealized uses and its realization in actual use. Most technologies are developed with certain uses in mind, targeting desirable characteristics or behaviors for their users, what Bruce and Rubin call use as *idealized* by its designers. What often happens during actual use, however, is a *realization* process, or a process by which users shape the technology to their own ends. If we are to take control over the medium and have it serve us, we must learn to recognize the ways in which we shape ourselves to fit the technologies we use.

Few instructors, however, question idealizations of ITV technology. J. M. Neff (1998) notes, "For faculty in most disciplines, televised instruction poses little overt difficulty because it supports traditional methods of delivering education—lecture, discussion, examination" (136). The ideal use of ITV, then, appears to be presentation-style delivery with the teacher positioned as expert and the students positioned as passive recipients of prepackaged knowledge. This ideal is further entrenched by norms and conventions associated with commercial television, especially newscasts: stories and notes are compiled, transformed into scripts, and read by experts, who sit behind desks with graphics displayed over their shoulders.

The problem with presentation-style delivery, as both Neff and Anson point out, is that current writing pedagogy does not usually follow this teaching model. Instead, we have found that "students learn well by reading and writing with each other, responding to each other's drafts, negotiating revisions, discussing ideas, sharing perspectives, and finding some level of trust as collaborators in their mutual development" (Anson 1999, 269). In other words, we engage in highly collaborative writing workshops. To some extent, the features of ITV technology can interfere in this dynamic: time for discussion is discrete, bodies are distributed across geography, exchanging drafts must take place through other media, and distant students feel silenced because local students see it as a "hassle" to interact with them.

All these obstacles—both perceived and real—to the pedagogy we prefer indicate that we must reinvent the norms associated with ITV technology and create a realized use of the technology that supports collaboration, exchange, and interaction. Open discussions can help students become more context sensitive, but allowing them to reinvent their classroom behaviors to take advantage of the strengths of the medium will do even more: it will allow them to become better communicators, more effective collaborators in their learning experiences, and more powerful agents for change. They will begin, in short, a process of realization that will shape the technology to the social situation they create in the classroom. The following techniques offer ways of beginning a healthy process of realization for ITV technology in use.

First, we suggest that participants reconfigure their room to reconfigure conduct. Most ITV classrooms are arranged lecture-hall style: students sit in rows in front of a "stage" area equipped with a podium and an overhead projector. The idealized use of these rooms has the instructor in the traditional place of authority, at the front of the room, moving only as far as her microphone cord or camera angle will allow. Students—local and distant—sit and take notes. These are not the only possibilities for conducting an ITV class, however, even if they are the ones idealized by the designers. We suggest that teachers recognize the symbolic, authoritative space the front of the room holds and then consciously work to share that space with students by inviting, even requiring, them to participate in that space.

For example, in one of our classes, we ask that students spend some time at the front of the room at least once during the course, even if that means simply assisting the instructor. Having students work in groups from the front is especially comforting because students often find strength in numbers. To facilitate group work, we advocate forming groups across sites and assigning roles to group members—such as summarizing content, leading discussion, conducting a workshop—so they can further negotiate the authority space according to content and gain experience in collaboration with distant team members. We've found the use of agendas, which indicate time as well as content responsibilities, to be particularly helpful because students know what is expected of them, can be prepared to contribute, and can consider the adjustments they'd like made in the technology to accommodate their needs. Whatever the activity that brings students to the front, teachers need to be careful to demystify the technology: explain how controls work, demonstrate what monitors show and why, assist with sound checks, and provide time for experimenting and learning in addition to presenting.

Along with sharing the authority space, teachers should share student seats as well. Discussion times can be led by students or conducted with no one at the front of the room, to reinforce the value of students' participation and contribution. These shifts in geography can produce shifts in roles and reinforce student agency. Also, defining students' roles and responsibilities can open up possibilities for using both the room and class time more productively. If students see themselves as contributing to the flow and content of the class, they tend to take more control over the technology and the space so that their contributions are recognized and valued by others in the class.

Helping students define roles and goals can create a workplace meeting atmosphere not uncommon to many professional situations, where teams of people work together to accomplish something for the common good. From a more practical (instrumental) point of view, students are given opportunities to explore and use the technology, to present materials and lead discussions, and to experiment with camera angles and overhead devices. From a pedagogical standpoint, we are engaging with students in student-centered course design and implementation. If we support these kinds of student-led activities, the technology can be realized in ways that recognize the social, rhetorical situation while exploiting the strengths of the technology itself.

Second, we suggest that instructors be attentive to the language all of us use to describe interactions and re-create it to foster productive communication. Partly because many instructors spend most, if not all, of their time at one site, remote students often feel isolated, which surfaces clearly in their language. They describe their contributions, when in the form of questions, as interruptions of the normal flow and often begin their interruptions by apologizing. The effect of interrupting is further heightened by the cameras' sudden activation, precipitating speakers' abrupt appearance on the monitors. Local students reinforce the negative perception by turning their heads to look at the change on the monitor, emphasizing the feelings of distraction. "Interruption," as an unarticulated feeling or as a voiced complaint, fosters neither feelings of individual worth nor classroom community. Validating distant students' desire (and right) to contribute to class becomes partly a matter of changing the language students use to refer to their own activities and to those of their classmates.

Other communication scenarios beg attention when working in an ITV classroom as well. "Normal" conversation, with its overlapping turn taking and spontaneous commentary, simply doesn't work with ITV technology. Sound delays fragment or truncate people's comments, monitors and cameras make it difficult to identify who is speaking, and speakers can and do talk over one another. Therefore, we must decide on new protocols—such as students' identifying their names and site location when they speak—to ensure that students' voices do not get lost in digital deflections and electronic voids. Sometimes instructors must act as traffic controllers, but they need to be careful not to act as conduits for communication. We found it helpful to insist that students hand off the conversation to each other and refer to the author of previous contributions by name, thus requiring them to learn each other's names and reinforcing understanding that people on monitors are people, not merely virtual representations. Seating charts, for both students and technicians, help facilitate turn taking and quick location of who is speaking. We must insist as often as possible that students talk to one another directly, rather than through us, to build the interaction often lamented as missing in ITV communication.

Third, we've found that simply building in time for students to talk with one another—get some "face time," as one student put it—can break students out of their passive reception mode and increase their feelings of belonging to a unified class working toward common goals. We like to leave about five minutes at the beginning of class for people to connect a little before we get down to business, and we try to give collaborative groups (especially cross-site groups) time near the end of class to work, plan, or just plain chat. Students might also request that the "transmit auto-mute" mode, which blocks sounds, be used during videos, discussions, or breaks. Just as students find it tedious to look at themselves constantly during class, they sometimes find it tiresome to consider every word before they say it, even during informal conversations with their local classmates. Turning off the sound for brief periods of time gives students needed breaks from feeling on-camera during class.

We often hear that technology dehumanizes students and instructors. If this is so, it is because we conduct ourselves in ways less human; in other words, we gear our responses to the technology and not to the humans it connects. To emphasize the human connection, we try to initiate off-topic chatter and laughter on occasion, point out to students across sites their common interests, and allow students to move around—sometimes out of the gaze of the camera—so that they aren't sitting still for long stretches of time, feeling under surveillance.

Finally, we suggest collaborating with students in exploiting the medium to get more out of it and developing media-rich contingency plans for times when the medium fails. One of the assumptions we can safely make about ITV—or, for that matter, about any communications technology—is that it will occasionally fail. Developing contingency plans is a valuable lesson for students, who will have to deal with some of these same technologies in their careers beyond the university. As Elizabeth Tebeaux (1989) points out, analytical skills and imagination are two qualities students need to develop to survive in technology-rich workplaces. Moreover, contingency plans situate agency with the users, rather than with the technology, turning substantive views of technology back on themselves.

One of our instructors registered her frustration with the fact that some of our distant sites have experienced equipment malfunctions that have prevented them from accessing class. As a preventive measure for technology-related access problems, this instructor and her students decided to make videotaping every class a consistent policy across sites. Even if remote students are unable to interact directly with classmates, they can still access the material and observe the interactions that took place in their absence. Using suggestions from her students, this same instructor has also learned to make adjustments in how she uses her course packets. Instead of filling them with readings, she includes lecture notes and other study aids to help students fill in any gaps they might experience in their ability to access class. Videotaping classes for later viewing and providing materials that will support students at a distance models proactive responses to technical difficulties. It also uses the medium's familiar commodity—information delivery—to its fullest.

Another suggestion for a backup plan is to assign note takers for every class period, one or two students responsible for taking class notes and posting them to an electronic bulletin board or class listserv. Even if the technology does not fail, students find these notes valuable study aids, in addition to the fact that they provide valuable experience in a practice common in our profession: to take, compile, format, and share notes from meetings. With a little guidance, students not only learn how to develop contingency plans for dealing with technological glitches, but they also learn about the implications of their decisions and the viability of their solutions—some of the more subtle layers of practical-rhetoric-as-conduct.

IMPLICATIONS AND FUTURE DIRECTIONS

Our approach answers Meyer and Bernhardt's (1998) call for a workplace literacy students need in order to understand and act upon social, organizational, and technological systems—to think critically and solve complex problems. ITV gives the option of true practice, not just the role-playing that Meyer and Bernhardt advocate, to have students engage in, rather than act out, "scenarios where communication is likely to be difficult or strained" and to explore in collaboration with instructors the issues of power in discourse: "Who does the speaking? When? And under what rules?" (93). ITV offers the opportunity to expand our as well as students' understanding of communication media and distance delivery. This skill is an important one as communication technologies continue to overlap and become integrated: consider such new technologies as Web TV, Internet videoconferencing, electronic collaborative white boards, and other emerging technologies whose impact we can barely imagine.

One of the most important and challenging implications of our approach is that we must participate in it, which means that we must examine our assumptions, reactions, and attitudes along with students'. As Stuart Selber and his colleagues (1997) advise in their discussion about collaboration in hypertext writing, "[I]t may be important to collapse the distinctions between writers and readers, to subtly dissolve notions of who owns particular parts of a collaborative text" (263). We encourage a similar collapse and dissolution between students and teachers as we together explore ITV—a dissolution more radical, even, than that espoused by current student-centered pedagogy. Our readers should understand that exploring ITV while in use offers both instructors and students new opportunities to understand communication and media and to develop skills and ways of thinking that are clearly marketable as we face ever expanding ways to communicate at a distance. We must be willing and able to develop our sensitivity to the fact that not only these technologies, but also our very discipline and practice, are grounded in a complex world where communication itself is not just transactional or instrumental—it is transformational. It will be a difficult challenge to apply self-conscious reflection and participatory agency to ourselves in such a dizzying proliferation of technologies and communicative strategies, but we must.

We recognize, too, that our approach seems to take time away from what we see as the "content" of our classes—the lessons we'd like students to learn as they write and revise. To this concern we answer that if media are not themselves "content," then what are they? We do not expect students to learn how to run the sound boards and cameras, but we do expect them to learn to ask for what they want and to know enough about the technology to communicate their needs to the technicians responsible for running it. We expect students to learn to analyze the communicative situation and, from the constraints and opportunities presented by this situation, to realize a code of conduct through which they can not only get the job done but also articulate the values of our professional community. In short, we expect them to develop and exercise a literacy of agency by participating actively in shaping the technologies through which they communicate, collaborate, and work.

3

BILINGUAL PROFESSIONAL WRITING

An Option for Success

Elaine Fredericksen

Professional writing instruction has changed considerably during the twentieth century as practitioners have worked to develop usable theories of technical and business communication. As a new century opens, United States demographics show Hispanics emerging as the fastest-growing minority, already overtaking African Americans as the largest U.S. minority group. Census data indicates that "the Hispanic school-age population is growing faster than any other group in the country" (Gehrke 2002, 4A). Concentrations of Hispanics remain high in coastal states and are growing at astonishing rates in the Midwest and South. This demographic shift suggests the need for educators to adapt, create, and grow further as the century turns.

The University of Texas at El Paso (UTEP) has developed a Bilingual Professional Writing Certificate program as a method for increasing opportunities for English and Spanish bilingual graduates in the fields of business and industry. This program can serve as a model for other institutions of higher education fortunate enough to boast large numbers of Hispanic students. The model also has possibilities for adaptation to other bilingual or multilingual situations.

Located on the U.S.-Mexico border, UTEP is a comprehensive public urban institution, a midsized commuter campus located in the world's largest binational metropolitan center. The student population ranges from 60 to 70 percent Hispanic, mainly Mexican American and natives of Mexico who cross the border daily to attend classes. Approximately 3 percent African Americans and 4 percent Asian American, Native American, or international students from non-Hispanic countries comprise the rest of the non-Anglo population. This university is well situated to try new programs that encourage the success of Spanish-background students as well as other students interested in becoming fluent in the two languages.

THE PROBLEM

Fluency in two languages should give students advantage in school and later as job seekers, but this has generally not been the case. One language tends to be dominant over the other in all students except the most totally bilingual, a rarity because dual proficiency requires the same degree of education in each language. Many Hispanics raised in the United States understand Spanish because they hear their parents, and particularly their grandparents, speak it regularly; however, they may not speak it well themselves and often do not know how to read or write it at all. Recent immigrants and Mexican nationals use Spanish as their dominant language and may have severe difficulties reading texts and producing technical or business documents in English.

The major manifestation of dual language backgrounds is that students speak and write with an accent. Rather than being an indication of a special talent and therefore an advantage, their accents hinder their classroom success and serve as an obstacle in their job search. Our program aims to change this situation and to make bilingualism an asset.

Stereotype threat presents one challenge for Hispanic learners. C. M. Steele (1997) explains *stereotype threat* as a situation wherein members of a minority group, even those who have been successful in school and at work, fear they will be labeled as inferior. Steele believes that "susceptibility to this threat derives not from internal doubts about their ability . . . but from their identification with the domain and the resulting concern they have about being stereotyped in it" (614). This stereotype means that even Hispanics graduating from secondary schools with high grade point averages may worry that their talents will not be recognized in college. They fear the stigma that "people like them" belong in low-paying jobs that require only high school diplomas. Students suffering from stereotype threat need reassurance and a sure sense that their talents have worth. L. I. Rendon (1994) argues that universities must validate students. Their aim should be "to remove obstacles to learning, to instill in students a sense of trust in their ability to learn, to liberate students to express themselves openly even in the face of uncertainty, and to know that the way they construct knowledge is as valid as the way others construct knowledge" (47). Our program attempts to validate students in these ways.

Traditional teaching methods represent another challenge that confronts Hispanics in higher education. Despite many pedagogical innovations over the last decades, most college and university classes still follow the lecture-exam pattern. Some classes include discussions, and many enlightened teachers require some group work. However, as pointed out in Goodlad and Keating (1994), "the needs of students whose cultural and ethnic backgrounds tend to be outside the traditional mainstream are typically not met by what might be characterized as one-size-fits- all education" (273). Technical writing classes, falling under the jurisdiction of different departments, have been somewhat slow to incorporate new instructional pedagogy. Certainly, the language of instruction and production tends to be English and only English. The model presented here suggests a new pedagogical approach through a linking of two languages.

THE PROGRAM

The Bilingual Professional Writing Certificate program at UTEP grew out of the frustrations of attempting to teach technical writing and business communications classes to junior-level students who suffer from second-language interference. Two professors, one in English and one in languages and linguistics, surveyed students about their interest in English/Spanish professional writing courses that would satisfy a university requirement and let them practice both languages. The response was overwhelmingly favorable, and the professors began planning a pilot program.

Fortuitously, the Fund for the Improvement of Post-Secondary Education (FIPSE), sponsored by the U.S. Department of Education, issued a call for grant proposals. The guidelines seemed a close match for the pilot. UTEP's successful grant proposal provided almost $223,000 to implement a three-year trial program.

The program structure is quite simple, requiring participants to complete two English courses and two translation courses and then to pass an exit exam. The entry course into the program is either a bilingual section of Technical Writing (English 3359) or a bilingual section of Business Communication (English 3355) in the Department of English or Introduction to Translation (Translation 3359) in the Department of Languages and Linguistics. One or another of the English courses is required for many majors or minors, including business, criminal justice, professional writing and rhetoric, computer information systems, marketing, and management. The translation class is the first course required for minors in translation. Thus, instructors have an opportunity to assess students' interest and expertise in bilingual situations when they are college juniors and to invite them to work toward a certificate. By taking an additional three classes in their final two and a half years, they can gain the benefits of certification.

Class requirements included the following courses:

1. Either English 3355 or 3359 (bilingual sections)
2. Translation 3359
3. English 4300 (This senior-level technical writing practicum reinforces students' abilities to write proposals and reports in both languages and places them in an unpaid internship with a local client.)
4. Either Translation 4381 (Business and Legal Translation) or Translation 4382 (Translation for Information Media) or Translation 4383 (Literary Translation)

Students must also successfully complete an exit exam that requires them to produce original documents (such as memos and letters) in both languages and to translate documents from English to Spanish and from Spanish to English.

Recruitment

Students are recruited into the program through introductory courses. We also distribute informational brochures campuswide, particularly utilizing advisors in the various related departments. During registration for each semester, we post large flyers around campus and hand small flyers to registrants. We also post bulletin broadcasts on student-accessed email. The school newspaper carries articles about our program several times a year.

Although advertising brings in students, the best recruiting tool seems to be word of mouth. Students who finish a course or two and find the experience useful recommend the program to others. Our bilingual technical and business communication classes always fill early in the registration process.

Student Profile

Student profiles have been gathered at the beginning and end of each bilingual section of English 3355 and 3359 throughout the three years of the program's existence. Academic majors varied for enrolled students, with the largest numbers in computer informational systems, marketing, management, and English. When questioned about their reasons for selecting the bilingual section, students reported that they were interested in both languages and that they felt good bilingual skills would be an advantage in achieving their professional goals.

Specific numbers vary from semester to semester, but in one representative survey, at the beginning of the semester 60.9 percent of enrolled students rated their fluency in conversational Spanish as excellent, and 26.1 percent rated these skills as adequate. In spoken English, 43.5 percent considered themselves excellent, and 34.8 percent rated themselves as adequate. Almost all had studied both English and Spanish in school for at least a year, but in terms of writing Spanish, only 26 percent felt capable of writing a short business letter or a technical report. Only 26 to 30 percent felt capable of producing these documents in English.

By semester's end, over 90 percent felt above adequate in speaking both languages. All students felt capable of writing a short business letter, and more than 80 percent felt comfortable writing a report in either language.

When asked what they had gained from taking the bilingual course, students in all semesters reported a positive experience:

"I actually learned how to write formal business letters."

"I gained a sense of team work and a better sense of technology."

"Technical writing skills. I learned how to focus on the audience."

"I have gained experience in writing in both languages. Also more confidence to write in a formal way."

"I've learned all the great ways to apply technology to school projects both in English and Spanish."

"A thorough understanding of the professional writer in the business environment."

The program now offers two bilingual sections per semester, with an enrollment of twenty-five each. All sections regularly enroll fully and early in the registration period.

Advantages

As most enrolled students seem to recognize, this program turns the perceived handicap of learning English as a second language into the advantage of being able to function effectively in two languages. Rather than thinking of themselves, or being thought of by potential employers, as inferior because they speak or write English as a second language, certificate holders may be selected over monolingual candidates because of their special training. This undergraduate program also holds promise for students who want to enter UTEP's new Ph.D. program in rhetoric and composition, which offers a unique bilingual option, allowing students with certificates to do graduate work in translation and cross-cultural rhetoric.

The program has the further advantage of highlighting business and technical exchanges with Latin America, one of the fastest growing and most neglected economic markets. Students who can communicate across cultures and language borders will find satisfying work while improving international relations.

THEORY

Strong theoretical support underlies our model for bilingual professional writing in the areas of intercultural communications, collaboration, audience response, and writing process theory.

Intercultural Communications

Technological advances in the twentieth century drew nations and cultures increasingly closer. Business and industries aiming for success in the twenty-first century must begin with the understanding that they operate within a global community unlike any before—a community demanding the fluent use of many languages. Perhaps more importantly, they need awareness of different cultures and how they do business and of the ability to hire employees who can comfortably cross physical and cultural borders. A look at statistical compilations shows that in 1997 about 335,000 Americans worked for foreign-owned companies either in the United States or overseas. More than 30,000 American companies exported goods to other countries. Even before NAFTA, the U.S. earned $33.3 billion a year from exports to Mexico alone.

Within the United States, growing numbers of Spanish-speaking buyers require advertising and marketing campaigns that differ from those designed for English speakers. They also need Spanish-language product user manuals, information sheets, assembly directions, and other documents. In the interest of safety and efficiency, multilanguage documents and signs have to be provided in workplaces that employ workers whose dominant language is Spanish. Obviously, employers need professional communicators prepared to take on the task of producing these materials.

Speaking, reading, and writing a second language go a long way toward helping employees serve the changing needs of U.S. companies in a global community. As Iris Varner (2000) points out, "If business partners do not speak a common language, the entire intercultural business communication approach will be influenced by the dynamics of interpreters" (48), which can slow down or otherwise hamper transactions. Furthermore, language skills must be supplemented with "insights into social behavior, attitudes toward morality, self-perception, and the role of hierarchy" (41). The effective cross-cultural communicator avoids stereotyping, while recognizing that members of a given culture share attitudes and practices that influence their response to outsiders. Corporate cultures differ even within the United States; in an international community, styles present much greater variety. A lack of information can cause embarrassment and loss of business, even in something as seemingly basic as setting and keeping a meeting time. Thus, educators interested in training professional writers must teach them the complexities of cultural difference.

Collaboration

Teamwork and collaboration have long been staples of U.S. business and technology. Today's workers must expand the ability to work with others within their own culture to an expertise in working with people from different cultures. Students learn to cooperate on the job by experiencing collaborative learning in the classroom. Furthermore, students learn more quickly, remember better what they have learned, and form stronger bonds of friendship and mutual respect with classmates and teachers when they take part in collaborative learning communities. Theorists (Johnson and Johnson 1988; Schmuck and Schmuck 1997; Slavin 1990) show that working on group projects in semester-long teams helps minority students (women, lower-class students, nonnative English speakers, and others) gain confidence in their abilities. This confidence helps in job interviews and creates independent workers.

Audience Response

When writers produce messages, they engage in a process of encoding. As they do this encoding, they assume that the recipient will be able to decode the message without significant difficulty and with a low degree of error. It is easy to see how important this can be if the message includes, for example, specifications for materials used in the construction of a bridge or calibrations of equipment for use in hospitals. Audience response theory concerns itself with the proper encoding of messages and with the importance of directing attention toward the proposed receiver.

Differences between senders and receivers can cause confusion within a given company, city, state, or country because no two people have the same life experience. Confusion or misunderstanding looms even larger when the sender lives in a different country from the recipient and larger yet when one speaks a different native language from another person. Clearly, anyone engaged in bicultural communication or translation must be aware that "when a message reaches the culture where it is to be decoded, it undergoes a transformation in which the influence of the decoding culture becomes a part of the message meaning" (Samovar and Porter 1988, 21). Bilingual professional writers have to think forward to the decoding process and anticipate problems caused by ambiguity or lack of accuracy. They may need to use what has been called "back translation" (having one writer translate the document in the second language and another writer translate it back into the original) to check for potential problems (Samovar and Porter 1988). They also need to decide when this rather time-consuming operation is necessary and when it is not. Such decisions require skill that can come only with practice.

Process Writing

The theory of writing as a process drives most college English composition classes and offers useful assistance to bilingual professional writers. Theorists (James Britton 1975; Lester Faigley et al. 1985; Donald Murray 1972; Janet Emig 1971; James Moffett 1968) and many others urge teachers to emphasize the way people write rather than focusing only on the final product. This matters a great deal in the field of technical communication where workers are accustomed to sending rapid email messages, memos, and letters as they conduct business. These communications in English, and even more in multilingual situations, may be poorly worded and ambiguous. Writing teachers can guard against these problems by encouraging students to think about what makes good writing. They can stress the value of putting a message aside briefly and then reexamining, revising, and correcting their writing before sending it. They can also urge writers and translators to submit work to an editor—a supervisor or coworker—to make sure it communicates effectively.

THEORY IN THE CLASSROOM AND BEYOND

Technical communication students do not study theory in isolation. They must have opportunities to put theory into practice. Our junior- and senior-level technical writing courses require students to do what they have read about. In their junior year, students work in classroom companies to complete teacher-directed projects or community projects for which their teams can produce written documents. In their senior year, students act as interns in U.S. or Mexican companies on either side of the border. Each of their projects includes creating original documents (manuals, brochures, newsletters, information sheets, and such) in two languages and/or translation of existing documents. Students at UTEP have worked for hospitals, clinics, nonprofit associations, and for-profit businesses and industries both in El Paso and in Ciudad Juarez. These activities create real or realistic situations where students work together producing a bilingual product. As they collaborate, they are challenged to achieve more than they have previously and are aided in this collaboration by other group members with more experience or more skill in a particular area. This challenge is especially effective in classes like ours where some students are more proficient in English and others more proficient in Spanish. The effect is a social and professional situation that replicates the working world and encourages cognition and creativity. By working and learning as a team, all the students develop greater skill in both languages as they also develop their professional writing skills.

QUESTIONS OF ETHICS

Essentialism

People from one culture who work with people from another culture must always guard against the dangers of essentialism, that is, the viewing of others as a group rather than as a collection of individuals. Professional writing educators have a responsibility to teach about cultural differences while warning students that not all members of a culture are the same. U.S. citizens certainly share certain qualities and understandings, but a young white woman raised in Mississippi differs in basic ways from an elderly black man from the same state. Nor is she much like a middle-aged Hispanic raised in New York City. These people speak the same language and pledge allegiance to the same flag, but they do not fit a simple pattern. Any attempt to categorize them as "typical Americans" will lead to stereotyping and essentialism.

How much more complex is it, then, to try to put labels on Latino Americans who do not share a federal government, a monetary system, or, in the case of Brazil, even a common language. Doing business in Latin America, or anywhere in the world, requires careful study and time in a target area because "the world does not divide into neat cultural packages that can be labeled, sorted, and inspected with ease" (Brake, Walker, and Walker 1995, 80). J. Leigh (1998) reminds us that learning another language brings the learner into another world, but the learner "must be able to look into, not at, the culture of the other" (90). Professional writers must not only know about the customs of other countries but also make allowances for individual differences among citizens from those countries.

Translation

The specialized work of translation presents a wide range of ethical considerations. Translators can, intentionally or otherwise, create "a screen between cultures and have the potential to apply only their meanings to the words spoken in either language" (Leigh 1998, 41). Thus, they can skew the reading of any given text. This skewing opens doors to power abuse, where the translator debases one culture or elevates another through subtle uses of language. People living and studying in the United States tend to recognize the country's many freedoms, its economic stability, and its international prominence. An incautious translation can show a preference for these qualities and the accompanying middle-class standards prevalent in the United States to the detriment of people from other countries who will read the translation.

According to L. Venuti (1998), translation is subjective and unscientific. Translation teachers must warn students to be as objective as possible and to avoid imposing their personal standards on the text and its future readers. Moreover, both teachers and students have an ethical responsibility to study and remain sensitive to social and cultural difference. The translation element of our program alerts students to the philosophical implications of navigating between languages and warns them to work carefully. Students also learn to submit their translations to ethical editors and to revise as necessary before seeing the translation into print.

Although some errors in translation can be amusing, others lose business, cost money, or endanger lives. A classic example of lost business is the well-known Chevrolet campaign to sell its Nova in Latin America. Someone should surely have realized that *no va* in Spanish means *it doesn't go*. Needless to say, the campaign did not sell many cars.

More dangerous are errors that stem from faulty understanding of vocabulary for measurements, materials, and other important specifications. Poorly translated manuals and assembly guides are annoying for consumers trying to assemble a desk or a bookcase; they can be fatal when technicians use them to assemble life-care systems or airplanes.

Bilingual professional writing programs offer excellent opportunities to teach ethics. Essentialism and translation should be discussed, but these also provide a strong introduction to more general issues of ethical practice in business and industry.

PROBLEMS AND TRANSITIONS

Faculty members of the Bilingual Professional Writing Certificate program and staff at UTEP continue to examine and modify practices. The program takes from two to three years to complete, and we find that too few students who come aboard in their junior year manage to finish all four required courses. Thus far, only one candidate has received a certificate. Others are close, but many students tend to go on to graduate school or find jobs—due partly, we believe, to the skills they gain in our courses—before finishing all certificate requirements. In an effort to get students interested in the program earlier and to plan all four classes into their last two years of undergraduate work, we will attempt to make sophomores more aware of the certificate and its benefits. We have also met with the deans from the colleges of business and engineering and have sent brochures to all faculty who advise students. We hope these faculty members will not only allow time in their class schedules for completion but also urge more students to enter the program.

Another problem that other institutions adopting our model will certainly encounter is the need for skilled bilingual faculty. Although we have been fortunate in hiring an English-Spanish bilingual professional writing teacher, we have also taught the class with a team approach, including an English professor in combination with one from languages and linguistics. We continue to train existing faculty to teach in our program, as demand for our courses increases. Flexibility should play a role in staffing certificate classes.

Currently, no textbook exists that meets the needs of a bilingual technical writing class. We have been using bits and pieces of available texts plus quantities of teacher-developed material. Elaine Fredericksen and Carol Lea Clark are writing a more suitable textbook that should be published within the next few years. This resource will include translation exercises, cultural notes, and a strong emphasis on relationships with Spanish-speakers within the United States and with Spanish-speaking countries around the world.

MODEL DISSEMINATION

We consider our program a moveable feast. Simplicity characterizes the model and makes it readily transportable to other institutions. The model also leaves room for adaptation to other target languages and to other regions. Our program focuses on two languages: English and Spanish. With qualified faculty, schools could include a number of languages in the same class, making it a multilingual rather than bilingual experience. Because our campus sits on the border between the U.S. and Mexico, students have ready access to businesses and industries working in two languages. In areas without such access, instruction can be supplemented with guest speakers, videoconferences, and email communication.

Although adopting any new program takes planning, we believe this model can be put into place with a minimum of preparation and little expense. Our experience suggests that the effort will be worthwhile.

CONCLUSIONS

As business and industry take on a more international flavor, the need for well-trained bilingual (or multilingual) professional communicators continues to grow. These personnel can be either English or other-language dominant but must have good skills in reading, speaking, and, especially, in writing both (or all) languages. J. Gilsdorf and D. Leonard (2001) point to the advantages of hiring nonnative English employees, "including exposure to new markets and suppliers, additional talent and skill, and lower costs" (441). Bilingual professional writers reduce dependence on costly translators, improve relations with non-English-speaking associates and customers, and eliminate embarrassing or dangerous communication errors. Educators have placed too little emphasis on these advantages and have often overlooked the potential of bilingual students as future technical and business writers. A certificate program like the one at UTEP can give these students the impetus they need to find jobs and succeed in a global business community.

This program offers particular promise in a world where technology has changed "national and international economics, demographics, and the structure of society" (Tebeaux 1989, 138). Our program prepares students to meet these changes in positive ways. Their specialized training equips them to work in the border community where they now live and also to move into other multicultural communities as the evolving workplace requires. Rather than continuing to be identified as an underrepresented and underemployed minority, these bilingual communicators will take their place as a vital element of the contemporary workforce.

4

DISCIPLINE-SPECIFIC INSTRUCTION IN TECHNICAL COMMUNICATION

W. J. Williamson
Philip Sweany

This chapter explores possibilities for pedagogical innovation offered by interdisciplinary ventures in professional development. Introductory technical communication courses offer their instructors and host departments opportunities to link conceptually and pedagogically with representatives from disciplines across campus. Too often, however, such opportunities are not acted upon. In this chapter, we discuss our participation in a two-year project that linked technical communication and computer science courses for students entering computer-related professions. We discuss the impetus for this venture, describe the linkage of courses that resulted, and explore the advantages, disadvantages, and challenges posed by the venture. We close the chapter by offering advice to others who may be interested in launching similar projects or redefining whole programs with an interdisciplinary focus.

Our final recommendations focus on two key elements of the problems posed by interdisciplinary collaboration: identifying institutional constraints and potential sources of support and engaging in thoughtful reflection on the limitations and expertise of the faculty who hope to bring such projects to fruition. We assume that most interdisciplinary ventures will, like ours, be limited to revising specific courses. However, the project of reimagining courses leads necessarily to consideration of full curricula and opens up at least the possibility for transforming entire programs.

THE IMPETUS TOWARD INTERDISCIPLINARY PEDAGOGY

Courses that introduce students to the fundamentals of technical or professional communication are staples of college-level curricula. Whether such courses are required components of a general education curriculum or offered as electives, they have two characteristics in common: many students do not look forward to taking them, and many faculty do not look forward to teaching them. We foster a different attitude toward the course. We see the introductory course in technical communication as an ideal site for connecting disciplines and emphasizing professional development.

Introductory courses in technical communication vary in content and strategy, but at their core, they focus in some way on the communication of information related to scientific, technical, or other special knowledge areas. Communicating information has been a perennial struggle for scientific and technical professionals. Likewise, science and technology are not the typical centers of expertise for disciplines strong in communication. Simple logic suggests that some mutual growth may result from any collaboration between these groups. Nonetheless, simple logic in this case defies institutional tradition. We believed that an interdisciplinary linkage of some sort could help students better understand both computer science and technical communication. And in the process of developing an interdisciplinary classroom venture, we hoped to learn more about our disciplinary differences and commonalities as well, especially with regard to pedagogy. As we show later in this chapter, interdisciplinarity emerged in this project in many ways, including the initial course design, the selection of a course project, the execution of our course design and, to some extent, the makeup of our project teams in the course itself. And although we recognize that the core concept of offering a discipline-specific course of this kind is not new, even at Michigan Tech's campus where we were both teaching at the time, we would argue that deep-rooted divisions among colleges, departments, and disciplines, coupled with short institutional memory for pedagogical innovation, makes projects such as ours innovative.

We agreed from the beginning that a course linkage between computer science and technical communication presented an opportunity to push students toward thinking about themselves less as students and more as developing professionals. Emphasizing professional development became our conceptual link between the subject matter of our courses. We argued that students may be able to succeed by compartmentalizing their knowledge into discrete structures (such as classes, tests), but working professionals could not. (Of course, we attempted to challenge the myth of separation in the academy as well, but that attempt is a discussion worthy of its own narrative.) The emphasis on professional development emerged in the way we presented the course to students, drew discussions back to issues of responsibility, responded to their work, and involved students in a project that asked them to serve a real client. Throughout, we addressed students as developing professionals and, in doing so, asked them to raise their expectations of themselves and their classmates to a professional level.

The time and energy we invested in this project was in part an attempt to address the negative attitudes students bring to their work in introductory technical communication classes. To some extent, we were successful in altering those attitudes. Students responded favorably in course evaluations, suggesting that we continue the practice of linking computer science to technical communication. Although many students commented that the linkage finally made the introductory course in technical communication relevant and worthwhile (and did not recognize as readily the importance of technical communication to their work in computer science), some students did understand and appreciate the full impact of our pedagogical venture. To understand the students' preconceived notions about technical communication, we need to frame the introductory courses institutionally.

Introduction to Technical Communication at Michigan Tech

The Department of Humanities at Michigan Tech typically offers six to eight sections of Introduction to Technical Communication (Humanities 333) each quarter. Although a few sections are taught by tenure-track faculty, most sections are taught by graduate teaching assistants who have demonstrated classroom expertise in first-year writing and who have some interest in teaching technical communication. Sections tend to fill quickly to their twenty-five- seat capacity due to general campus demand; most majors on MTU's campus require the Introduction to Technical Communication course, and all majors can take the course as an elective. MTU students typically enroll in the university's general education technical communication course during the final year of their undergraduate curricula, when they are concurrently enrolled in upper-division courses in their majors. Students are not usually encouraged to overlap these experiences in any way. As a result, the introductory technical communication course is often viewed as the final writing course that students need "to get out of the way" before they can graduate.

The course is housed in Michigan Tech's Department of Humanities, which includes faculty from a variety of disciplines, including rhetoric, technical communication, literature, composition, philosophy, and modern languages. The department's disposition toward Humanities 333 has varied somewhat in recent history. At times, the course has been considered little more than a service course in the most limiting sense. At other times the course has generated interest from faculty and graduate students alike as a site for interdisciplinary investigation of communication practice. For a time, MTU's undergraduate Scientific and Technical Communication program was directed by a faculty member whose expertise lay outside traditional technical communication studies. In many ways, this administrator's direction introduced invigorating and transformational ideas into the overall conception of the undergraduate program. However, this administrator also questioned the significance of investing faculty time into supporting campuswide, service-oriented courses in technical communication. As a result, the Introduction to Technical Communication course was not formally administered by any subgroup within the department. Although this hands-off policy changed under the guidance of the following administrator, several years later the department still feels the lingering effects of that decision and has been forced to retool the course because it has been reclaimed into the department's teaching culture.

Despite documented early successes in writing-across-the-disciplines initiatives at Michigan Tech (see Young and Fulwiler 1986, ; Elizabeth Flynn et al. 1990), Humanities 333 has always been powerfully influenced (not always positively) by pedagogical tradition and institutionally imposed limitations. For example, due to the high concentration of engineering majors enrolled in these courses, students tend to address engineering concerns more than those of any other discipline on campus. Nevertheless, engineering and science faculty at Michigan Tech have expressed concern that Humanities 333 is too broadly defined, asking for courses that serve the particular needs of specific disciplines. Our interdisciplinary venture grew out of an initiative from computer science faculty who shared such concerns.

In the next part of our discussion, we outline the moves that took us toward establishing curricular links between computer science and technical communication at MTU.

A Conceptual Framework for Course Linkages

Our interdisciplinary venture was based on the assumption that linking introductory courses in technical communication directly to the professional development programs of other curricula seems a logical move toward more effective pedagogy and more fruitful relationships among disciplines and departments. We also felt that this linkage had potential for addressing more-general workplace concerns about communication abilities among college graduates: by linking communication to disciplinary content, we hoped to spark greater interest in the subject matter and to make it more central to professional development in computer science.

Certain evidence suggests that such curricular relationships have been beneficial for both students and faculty. For example, at Georgia Tech, mechanical engineering students are introduced to technical communication by writing faculty, who teach communication within the context of engineering courses (Donnell, Petraglia-Bahri, and Gable 1999). At the undergraduate level, technical communication is taught in four courses; the curriculum design emphasizes the simultaneous development of communication ability and professional engineering expertise (114). At the graduate level, students are returned to a novice level as professional communicators, at least with respect to their transition to the role of researchers; communication is linked explicitly to professional development in a series of courses and seminars (115–116). At Worcester Polytechnic Institute, John Trimbur (1997) teaches a course called "Writing about Disease and Public Health" to students from primarily biological and technical communication backgrounds. In that course, Trimbur asks students to evaluate the transformation of medical information in its travels from the academy to public spheres. For Trimbur and his students, understanding the delivery of information is critical to professional activity and responsibility in biological professions. The course emphasizes the awareness of professional responsibility in the presentation of information to a variety of audiences. Although these arrangements do not involve specific course linkages, as ours does, the overall motivation and impact for each course is much the same, emphasizing communication in the context of professional development.

There are good examples of course linkages as well. Brian Turner and Judith Kearns, for instance, describe the linkage between a first-year composition course and a first-year history course at the University of Winnipeg (Turner and Kearns 1996). This linkage was designed to help students adjust to the rhetorical demands of college-level writing and to help them understand the content of their history course (5). Dennis Lynch (1997), a Michigan Tech colleague, is involved with another first-year course linkage, this one between composition and biology. Among the goals of the linked courses, Lynch identifies the study of how biologists write and communicate and what it means to become a biologist (163). In each of these course linkages, as with the examples of the conceptual and pedagogical links discussed previously, there is a sense and a hope that by introducing interdisciplinarity into science, history, engineering, and communication instruction, there will result a greater understanding of each element. In effect, the hope is that the whole really will be greater than the sum of its parts.

Among the efforts we highlight here, the consistent element to developing interdisciplinary linkages is the emphasis on professional development. That emphasis is central to our work as well. From the first, we felt that when framed as courses in professional development rather than as a series of exercises about writing forms or the mechanistic development of text, technical communication courses can become explorations of what it means to become a responsible, practicing professional. Thus, we presented communication as the social medium within which products such as software packages develop. This richer focus on communication as a social activity is likely to be more interesting and useful to people across the disciplines. It is also likely to be more rewarding for students. In the context of computer science, course linkages open opportunities to discuss the relationships in professional life among communication, product development, and customer service. We can open up opportunities to engage faculty and students in discussions about responsibility in professional life, including the ways communication knowledge provides a medium for other professional responsibility. These courses also provide opportunities for teachers to talk about the differing roles professionals may play with regard to communication, including the different ways engineering, science, business, service, or communication professionals might deal with the same body of information in different contexts.

The theme of professional development drove us to design a seamless curriculum whenever possible; that is, we wanted our two courses to feel like a single course. Therefore, instead of treating the courses as two separate but linked courses, we developed the curriculum as a comprehensive, six-credit instructional unit. In the early going, we both met with both classes. Although we maintained final instructional authority in our respective courses, we approached their design as if we were developing a single course. As a result, computer science and communication interests and issues were as fully articulated as we could make them across the two courses. We were driven in our general design by two concerns: (1) that we foreground the importance of communication in both courses, rather than encourage students to separate computing and communication; and (2) that we provide students with a "real" project that would challenge them in ways that an invented project would not. We drew on the guidelines for successful writing-across-the-curriculum initiatives offered in Toby Fulwiler's (1991) "The Quiet and Insistent Revolution." We engaged collaborative learning groups in open-ended assignments that posed real-world challenges. We addressed student writing as managers rather than as teachers, offering guidance rather than grade-oriented commentary. We shared our values as communicators, researchers, and educators by discussing our pedagogical and research goals directly with students (183–185). We also looked to service-learning scholarship for assistance in drafting our specific project goals. For example, in "Technical Communication and Service Learning," Randy Brooks (1995) suggests that "the most valuable service learning includes reciprocity of outcomes: (1) the doing helps the community solve problems or address needs, and (2) the thinking helps the student develop disciplinary skills, community responsibility *(ethos),* awareness of cultural diversity through the integration of theory and practice" (12).

Our planning culminated in the pilot course linkage, offered in the fall quarter of 1997. In the next section, we describe that venture in greater detail.

SOFTWARE DEVELOPMENT IN LINKED COMPUTER SCIENCE AND HUMANITIES COURSES

During the spring of 1997, Philip Sweany, who was a member of MTU's computer science department at the time, initiated discussion about a curriculum revision regarding technical communication instruction. In the wake of those discussions, we began developing a course linkage that brought together the introductory course in technical communication with an upper-division software management course. Students enrolled in the linked courses worked to develop educational software for a local middle-school math class. In one ten-week quarter, we asked students to design, develop, and document a prototype software package suitable for early- development field testing. Students worked in teams of four to six people to complete several project-related items:

software package
user manual
design document (software design proposal)
functional description (description of software capabilities)
documentation plan
software testing documents
software maintenance plan
several progress reports

Only a few of the project-related tasks were completed for exclusive credit in one course or the other. Most course products received developmental feedback and grades from both instructors. When we responded to projects, especially written documents, we tried to treat the encounters as shop scenarios; that is, we responded to students' work less as teachers and more as managers. And in keeping with the shop atmosphere, students showcased their work in a "software fair" held in an open computing facility at the end of the quarter.

We prepared students as well as possible within a tight time frame to undertake their projects. As part of the learning and planning process, we asked students to evaluate existing educational software packages and their accompanying documentation. During the early part of the quarter, we connected students with subject matter experts from three areas: educational software design, small-group dynamics, and middle-school mathematics pedagogy. A faculty member from MTU's education department led a series of discussions on educational software design, including special focus sessions on developing interactivity and appealing to young audiences. A graduate student from MTU's humanities department led a series of discussions on small-group dynamics, including sessions on roles and role-playing and conflict management. Throughout the quarter, students were in contact with the math teacher who served as the project client via the World Wide Web and email. We gathered questions from students about her teaching methods, the teaching curriculum she used in her classes, and her students' experience with computers. She responded by posting information to a special section of her Web site devoted to the project.

Throughout the quarter, project-related discussions spanned issues in software and interface design, teaching and learning strategies, usability testing, and document design. We encouraged students to engage the theories they encountered in class discussions and in professional literature and, whenever possible, to extend those theories to their work. Although the software design project was the centerpiece of this linked curriculum, students participated in a variety of discussions and assignments that helped them develop the expertise they needed to complete their work.

Outcomes of the Software Design Projects

Because we did not initially have a long-term commitment to the project from our department-level administrators, we did not think it wise to plan to make the class projects an ongoing effort. That is, we insisted that project groups attempt to develop, test, and deliver complete educational software packages in eleven weeks. Given this tight time frame for software development, students were forced to focus on simple design plans and to establish relatively modest goals for the complexity of interactivity they designed into their software and manuals.

Predictably, students' attempts to design software for a middle-school audience were either too simplistic or too complex. Michigan Tech students (especially those enrolled in scientific and technological disciplines) tend to be adept at solving math problems. It was difficult for them to understand the relative simplicity of middle-school math curricula. Nevertheless, most groups managed to develop projects that the client said she would be willing to field test. Most of the software packages that resulted from that first run of the linked courses focused on presenting students with information from the curriculum and then quizzing them on the knowledge they were supposed to gain. In preliminary, informal field testing, students in the client's classes were often overwhelmed with the complexity of the problems presented in students' software. However, they responded favorably to the idea of integrating computers into their curriculum for any purpose.

One project group continued its work in an independent study that extended beyond the initial quarter. That group developed a game,

Lemonade Stand, that presented users with the problem of maintaining a successful beverage stand in the face of daily weather changes and fickle market demand for their product. To add an element of drama, the project group allowed users to avoid natural and social disasters (such as tornados and bullies) by solving additional math problems. This project group focused more than others on interface design and developing interactivity.

The user manuals that resulted from the software development projects were in some ways better developed than the packages themselves and with good reason. In a sense, the documentation drove the development of the software. Project groups submitted detailed written plans for their software and manuals well in advance of the completed projects. Those plans formed the basis for all of their future development. Because of the tight time frame, groups tended to stick to their original designs whenever possible so they would not have to go back and rework their documentation to fit new project developments. Thus, project groups kept to their general plans, focusing their problem-solving innovations on methods for accomplishing their original design goals. In the end, there were still last minute software changes not properly documented, but these instances were few. Overall, the user manuals were targeted more appropriately to the focus audience than the software packages themselves. Although students tended to assume that their audiences would find the software intuitive in its design, they also responded well in general to our suggestions for developing the documentation further.

Perhaps more significant than student successes in terms of their learning experience, however, was their recognition that they had all misjudged their audience in some way. The software fair was a showcase for what students learned and what they could have done, as much as it was an opportunity for students to discuss the software they actually developed. Few students were satisfied with what they had accomplished during the quarter, although most believed that their experience was not adequately represented in their final products. Unfortunately, MTU's quarter system left us with little opportunity to follow up on and encourage detailed reflection from software groups once the course ended. For many students, we could only hope that we had established a solid groundwork for such reflection and that they would engage in that process independent of their classroom experience.

We stated earlier that we believed students would benefit more from both their communication and software design courses if they were able to make connections between courses previously disconnected and to work with a real audience and purpose. And as hoped, students understood the relationship between their technical and communication responsibilities more fully in both classes than either of us had experienced prior to linking them. For example, in previous software development courses, Sweany observed discussions among groups primarily about technical issues in software development. In our linked courses, project groups discussed meeting user needs as much as they discussed solving technical problems. They argued about the logistics and merits of usability testing for both the software and its accompanying documentation. Teams in general were more concerned from the beginning about the way all the elements of the problem affected software design decisions (including user needs, information design, content, teaching strategies, learning styles).

In the end-of-the-term evaluations, students expressed a general support for the course linkage. In addition to receiving good marks in the numerically based evaluations, the course received positive qualitative comments:

> The format provided an interesting and relevant way to learn the material. It gave everything a sense of connection and direction.

> I would definitely recommend this course. I would warn them that it does require a lot of work, but that they will learn a lot.

> During the first week of classes, I considered dropping. It seemed that the work would be hell. At times I came close but I'm glad I stuck it out. I found that the project was fun and a good learning experience.

These early successes not only gave us many things to think about but also left us hopeful that we were on the right track, working toward a useful, productive goal that satisfied students, administrators, and our colleagues in teaching.

Despite rich possibilities for professional and curricular success, we still faced political and financial challenges in our endeavor to reinvent the introductory technical communication service course at MTU. We found that institutional momentum does not easily shift to support customized interdisciplinary education. In the next section, we reflect on some of the advantages, disadvantages, and challenges to establishing linked, interdisciplinary courses.

ADVANTAGES, DISADVANTAGES, AND CHALLENGES OF INTERDISCIPLINARY PEDAGOGY

Although we believe this interdisciplinary course linkage was generally successful, it is nonetheless a venture rich with complications. In this section, we address some of the advantages, disadvantages, and challenges we have come to connect with our efforts. Although we believe that this interdisciplinary effort ought to be adopted more widely, we understand why such curricular innovations, without careful planning and energetic self-promotion in the early stages, are likely to remain isolated experiments in pedagogical design. Because we are aware of the benefits as well as the challenges for implementing this curriculum, we address the issue in our final recommendations.

Advantages of Interdisciplinary Pedagogy

Our initial attempt to link courses in computer science and technical communication was met by students with increased commitment to connecting communication and computer science in their thinking about professional development. We have gathered feedback through a variety of means, including standard course evaluations from each course, anonymous questionnaires, informal interviews with students, and word of mouth. And some students have gone on to use their course projects as professional portfolio material in the job market. The linked-course project has begun to acquire a favorable reputation among first- and second-year students, many of whom now look forward to participating in the project. From the outset, we focused on four aspects of the course linkage we felt would most likely serve student interests and needs: the advantage of a comprehensive curriculum; professionally contextualized communication instruction; real-client motivation; and interdisciplinary exchange between faculty.

Comprehensive Curriculum Design

We have argued here that the linkage of courses in other disciplines to the introductory technical communication course could in the best of circumstances result in a comprehensive learning experience for students. Further, linked courses offer an opportunity to focus in greater depth on the relationships among subject matter expertise, communication activities, and responsibility against a backdrop of professional development. We still see this advantage as the overarching benefit for such curricular innovation.

Professionally Contextualized Communication Instruction

A consequence of linked courses is the opportunity to contextualize communication instruction with the acquisition of subject matter expertise in scientific and technological disciplines. This contextualization is more than a side effect of such pedagogical efforts; it is a real advantage to be pursued in course design and execution. We agree with students that contextualized communication instruction is far more relevant to their professional development than a course offered to a general audience of students who represent a random sampling of campus disciplines and departments. When instructors have the luxury of making technical communication courses immediately relevant to students, there is far greater potential for meeting everyone's expectations in the course.

Real-Client and Real-Project Motivation

Although it is not a direct requirement that a comprehensive, contextualized venture such as ours be anchored by a real project with a real client, we recognized immediately that the cumulative contact time gave us the opportunity to introduce this element to the courses with some hope of success. The advantage of serving a real client is an extension of the other circumstances of our venture; nonetheless, we would recommend that anyone interested in forging such linkages pursue the possibility of connecting to a real project. Again, this helps students contextualize their work more easily in a framework of professional development and offers them added motivation to succeed. A client will see their work in the end.

We recognized the extra effort required from our client to participate in a project of our devising. To benefit from the project, she had to invest time in helping students understand her needs and the needs of her students, as well as her curriculum structure and teaching methods, and she had to make time to test prospective software packages once they were ready for their audiences. We worked hard to ensure that her efforts were rewarded regardless of the success of our project. In return for our client's participation, we offered technical support for the delivery and use of the software developed by students, the promise of revisions to potentially lucrative software packages, and the possibility of new computers from future project funding.

Interdisciplinary Exchange between Faculty

Regardless of the classroom success or failure of a project such as this one, we feel strongly that one of the benefits of the venture, both for ourselves and students, is simple collegial exchange. To make clear our scholarly, pedagogical, and professional assumptions and goals, we had to understand them ourselves and be able to articulate them effectively. A significant element of our exchange revolved around tracing the intellectual legacy of our practices in response to the most elegant and confounding question of all: why? By going into our courses with a greater understanding of the values each of us held, we were able to anticipate how to best complement one another and when to use our professional differences constructively and creatively to spark discussion with and among students. Such pedagogical acts require trust that comes only from contextualized interdisciplinary discourse.

Although we have discussed these advantages already in this chapter, we have not addressed in any detail the drawbacks and challenges to interdisciplinary course linkages. We do that in the next two sections.

Disadvantages of Interdisciplinary Pedagogy

We would be remiss if we did not discuss the downside of interdisciplinary ventures such as the one in which we engaged. Although we discuss only two such disadvantages here, we round out the discussion by addressing challenges in the section that follows. Here we discuss problems of time investment in curriculum development and the negative consequences of extended student-teacher contact time.

Interdisciplinarity Demands Significant Time Investment

A course linkage such as this one demands a far greater investment of time than a standard university course preparation. Our discussions began several months prior to the first class meeting. During that time, we invested time in discussions about big-picture conceptual issues, logistical details, textbooks, guest speakers, assignment development, and other issues, including funding sources. Some of our efforts were supported by outside funding. Grants provided an alternate means of accounting for our time investment. Even so, this venture represents significant allocation of time above and beyond our typical course development.

Extended Student-Teacher Contact Can Have Negative Consequences

We already said that the extended contact time with students was one of the advantages of a comprehensive course linkage. However, that extended contact time can also become a disadvantage for both students and faculty. Students who struggle for any reason struggle in two courses, not one. If their struggles are confined to their peace of mind in the courses, then the disadvantage is personal to them. However, such is rarely the case, especially in a team-driven, project-oriented course. The course linkage represents a significant investment for students as well, and any problems are thus magnified. Time can be an opportunity or a prison.

Challenges of Interdisciplinary Pedagogy

In addition to the specific disadvantages of interdisciplinary course linkages, there are several obstacles to be overcome both to establish and maintain the venture. We classify them as challenges rather than disadvantages because they refer more to institutional difficulties than to pedagogical drawbacks. Each is also a call to respond innovatively to a dynamic professional context. We discuss five such challenges here: institutional obstacles; institutionalized disciplinarity; inconsistent rewards; funding; and sustaining commitment. After raising these challenges for consideration, we discuss several strategies for meeting such challenges and launching successful interdisciplinary teaching ventures.

Institutional Logistics Create Obstacles

Institutions do not always respond favorably to special limitations placed on course enrollments. Standard course prerequisites are easily accounted for by scheduling personnel and software, but our course linkage created problems for the campus registrar. The introductory course in technical communication is in high demand at MTU, and students from other disciplines were frustrated that some of the sections offered were closed to people outside computer science. The registrar's office did not share our enthusiasm for the linkage experiment and so resisted our efforts to have special limits placed on our sections.

Institutionalized Disciplinarity

Universities seem to be set up to encourage disciplinary isolationism, not interdisciplinarity, especially when working with undergraduate students. Students are required to take courses that range across several departments and disciplines as part of their studies, but rarely are individual courses cotaught by faculty from different departments or disciplines. We explored course linkages rather than a single cotaught course in part because we could not find a way to satisfy the university's need to assign credit for the course offerings to an individual faculty member.

Further, university faculty do not tend to encourage interdisciplinary thinking in their classes. Thus, students are not generally encouraged to build connections among courses from the complete range of disciplines they encounter during their studies. Institutionalized disciplinarity results in struggles to link courses and to get students to make connections between even similar subjects from different courses. Students are not trained to make interdisciplinary connections in their course work and thus struggle to do so when asked.

Inconsistent Rewards

Pedagogy is not scholastically significant in all disciplines. Rewards for pedagogical innovation are therefore uneven. Incentives for such activities are also uneven. Engineering and science disciplines do not value pedagogical research as highly as industrially connected forms of research. Consequently, interdisciplinarity may not be rewarded for those involved, if their tenure areas are narrowly defined. English faculty may not earn any returns for publications focused on pedagogy or on interdisciplinary studies. In some cases, this obstacle may be a significant barrier presenting real concerns, especially for tenure-track faculty.

Funding

Universities are not well set up to accommodate special allocations of resources. That is, when a special course is offered, such as Technical Writing for Computer Science Professionals, questions arise about who will accept financial responsibility for the course. Should responsibility remain with the department that hosts the faculty member or with the department whose special interests are being served? In our case, compromise was necessary. In university settings, funding allocations are highly political, and the issue of who pays is one best addressed early and with as much awareness of institutional politics as possible. We explored sources inside the university—through conversations with department chairs, deans of colleges, and upper-level administration—as well as outside sources for funding. Although we were able to secure sufficient support in the short run to launch the project, our results were mixed overall.

Sustaining Commitment

Sustaining interdisciplinary commitment to a project such as this one is difficult. Faculty in English or humanities departments may not feel qualified to teach general courses in technical communication. It is an added burden (and source of insecurity) to ask faculty to teach such courses to a single discipline. On the other side, not all members of technical or scientific departments are likely to support the interdisciplinary effort we embarked on. Neither are all faculty from either side of these relationships likely to see the benefit of contextualized communication instruction. Thus, personnel issues are a problem with any discipline-specific pedagogy because it is difficult to sustain commitment. Once a team commits to a venture, as we did, there may be no one else in either department interested in or willing to continue the project. Thus, a good idea, a successful venture may fail ultimately due to inconsistent commitments to the idea.

Despite the drawbacks and challenges we present here, we remain firmly committed to the idea of interdisciplinary course linkages. Thus, we provide some suggestions in the next section for increasing the likelihood of success in such ventures.

DEVELOPING SUCCESSFUL INTERDISCIPLINARY COURSE LINKAGES

Different strategies for sustaining interdisciplinary education will likely work better at different institutions, but there are a few things that we have found at Michigan Tech that seem to give us plenty of room for rethinking our relationships to other disciplines, while also rethinking the goals and strategies driving our introductory technical communication courses. To be of use to all stakeholders involved (students, faculty, administrators), there must be professional benefits from participating in the project. In rethinking the role of the technical communication service course for computer science majors, we engaged in several months of exploratory discussions, following a plan that might best be described by M. Jimmie Killingsworth (1997) as "pragmatic," focused in local, immediate, and serviceable needs (244–245). Although the specifics of our model may not work for all curricula, the process should be of interest to people looking for strategies to strengthen their present scientific, technological, or communication pedagogy, at either individual or programmatic levels.

In "Linking Communication and Software Design Courses for Professional Development in Computer Science," we (1999) attributed our success to seven strategies we adopted and that we think might be helpful to others who embark on similar ventures:

Plan curriculum development time
Plan faculty development time
Find a real client and project

Visit each other's classrooms
Plan an ongoing project cycle
Promote departmental consistency
Engage in vigorous self-promotion

Plan Curriculum Development Time

Prior to entering the classroom, we invested significant time to discussing our individual goals, project goals, and pedagogical values. We discussed external funding sources and possible project clients and how we might approach them. We outlined course syllabi, ideas for inviting guest speakers, discussion topics, and readings. We discussed the linkages between assignments, strategies for responding to student work, and creation of opportunities for encouraging client input. For our professional development, this time produced some of our most valuable and rewarding efforts. Our vision of the project emerged from hours of negotiations about the general approaches pedagogical and managerial to our courses. In the process of discussing what we could and would do together, we were forced to lay out our pedagogical theories and practices and work on crafting a compatible whole. Our discussions about the courses and the overall project helped us define and refine our teaching and researching roles and goals, including our personal, professional, and pedagogical goals, as well as our individual and collective measures of success.

Plan Faculty Development Time

We invested significant time early in the project developing shared expertise in a variety of project-related issues, including educational software design, service-learning design, Java programming, and collaboration. Each area played its part in preparing us to enter into the project as a teaching team. Critical in these discussions was the ongoing process of assignment expertise. We did not want students to feel as if any one person involved in the project had exclusive expertise; we wanted to present as much as possible that our ideas, goals, and expertise with different project elements were complementary and overlapping. For example, we did not want to project a sense that the technical writing activities were secondary to the success of the overall project. But neither did we want to project that only technical communicators understand sound communication principles. We fostered a spirit of collaboration as much as possible in our development and wanted to portray that same spirit to and for students.

Find a Real Client and Project

This strategy seems obvious, but projects can really vary. Although we focused on educational software, anything that gives students the opportunities to apply their talent and knowledge while helping the community can create a more enthusiastic work environment. Even simple projects can promote this commitment.

Clients themselves may come from a variety of settings—anyone who needs a project done or a problem addressed is a potential client. In our case, we set out to connect with another educator, and the local middle schools and high schools provided us with a long list of potential clients. In other courses where we have engaged in similar relationships with clients, we have sought out nonprofit organizations, community service organizations, and other faculty and staff from educational institutions. The difficulty is finding a client who recognizes that the success of projects often relies at least in part on the continued participation of the client. Absentee clients are too often surprised by the end result of student work. Clients who contribute to the success of projects without attempting to control them are more often satisfied with the results of their participation in service-learning environments.

Visit Each Other's Classrooms

For the first part of the term, we were regular participants in each other's courses. This participation helped promote a spirit of collaboration and connectedness we felt was important to display to students. They take the courses more seriously knowing that we do too. As the courses progressed, we held regular faculty meetings to keep each other apprised of developments in our classes as well as to share observations of student performance. We were careful to communicate our collaborations frequently so students would get the sense that the collaboration continued even after the start of the semester.

Plan an Ongoing Project Cycle

In this experiment, we carried only one project beyond the scope of the first quarter. We recruited one project group with a promising software concept and package to continue their work in an independent study. Ideally, we would have given every group the opportunity to work independently, but in this case this was not practical. Another possible source of project continuity would be to assign projects from one set of courses to groups in later sections of the same courses, asking them to contribute to the ongoing development of concept, software design, and documentation.

Promote Departmental Consistency

We promoted this project in our home departments to encourage other faculty who teach these courses to either adopt our approach or promote similar pedagogical values. Our hope was that we could recruit others to participate and thus ensure that even without our direct participation the concept of our pedagogy could continue. It was our hope to create a model for others to follow, to offer an opportunity to reconsider the strengths and weaknesses of the whole computer science and technical communication curricula. Conversation is healthy, we feel, and well-considered change can be healthy as well. However, this effort is an ongoing struggle.

Engage in Vigorous Self-Promotion

Part of our attempt to recruit others to the cause—other teachers, researchers, students, administrators, or funding sources—came in the form of self-promotion. We talked about the project with everyone, in meetings, in hallways, and via email. We submitted conference papers and publications. We submitted proposals for additional funding. Vigorous self-promotion is key to the success of any project and is particularly important to projects that ask people to rethink, revise, or reenvision their work if they are going to participate. Potential movement of any kind starts small and is sustained only by communication.

We recognize that interdisciplinary course linkages such as ours are not likely to become the model curriculum for computer science, for technical communication, or for professional development curricula. Unfortunate as this might be, we still feel the need to struggle to create and sustain an institutional atmosphere where such ventures are at least welcome experiments. That struggle opens up a trio of challenges for continuing investigation of questions beyond those that grow out of the discussion we have presented to this point. First, as a professional teaching community, we need to reinvent our pedagogy to fit the potential of new workplace relationships among professions. Nonacademic workplaces have often seemed more interdisciplinary than academic ones. Many projects in industry require the contribution of many disciplines and professions. Education ought to be a more interdisciplinary venture. But significant challenges remain before interdisciplinary education becomes an institutional reality. We need to identify those barriers and devise strategies for addressing them. Second, we need to create interdisciplinary places for discussion where values can come together, sometimes to collaborate, sometimes to complement, sometimes to conflict. Some conferences provide opportunities for interdisciplinary discussions, but too often our opportunities to engage with colleagues from other departments and disciplines are passed up. Certainly, our disciplinary barriers are in part a product of our disciplinary isolationism in curriculum design. Third, program administrators need to consider the short- and long-term ramifications of encouraging this interdisciplinary venture. In the short term, such projects open up pathways for focused reflection on pedagogical goals and practices. In the long term, such reflection can result in significant rewards. However, someone has to decide what resources, and in what amounts, are appropriate to commit to this project and whom among the many colleagues available it would be best to approach with this project. Again, how can we address this seemingly enormous set of challenges? This question bears further investigation, but we suggest some beginning places here in this chapter.

Interdisciplinary pedagogy of the kind we describe here is not new, but it is innovative. The present structure of colleges and universities does not make interdisciplinary work easy to develop or to succeed with. We met this challenge in the short run with hard work and hope for serving students well. Our early successes and failures provide fuel for future exploration in our work. And we end this discussion knowing that if teachers of technical communication and other disciplines can manage the potentially difficult logistics of a venture such as ours, there are real benefits for both students and faculty.

ACKNOWLEDGMENTS

The authors would like to acknowledge the contributions to this project of Fritz Erickson, Margaret Falersweany, Karla Kitalong, and Jennifer Williamson.

5

TECHNICAL WRITING, SERVICE LEARNING, AND A REARTICULATION OF RESEARCH, TEACHING, AND SERVICE

Jeffrey T. Grabill

Tensions among research, teaching, and service are real, and they are unproductive when they limit the type of work valued by the university (see Sosnoski 1994). There have been some notable attempts to rethink the work of the university and establish new ways to value a range of faculty initiatives that don't fit into the hierarchy of research, teaching, and service (for example, Boyer 1997). One of the more interesting attempts is the 1996 report by the MLA Commission on Professional Service, which takes as one of its starting places the imbalance among research, teaching, and service. The commission notes that service in particular is almost completely ignored or seen as an activity lacking "substantive idea content and significance" (171). There is nothing new either in the university's hierarchy of values or in the denigration of service. Yet this taxonomy of faculty work should be disconcerting to those of us who believe that a university must have long-term commitments to serve the community in which it is situated. But perhaps more problematic is the view of service as an intellectual wasteland.

My most general concern in this chapter is this view of service as lacking substance and significance. (I will focus, however, on community service learning rather than departmental or university service.) To be sure, the MLA Commission on Professional Service offers an intriguing rearticulation of research, teaching, and service into "intellectual work" and "academic and professional citizenship," with research, teaching, and service recast as sites of activity that can be found in both categories. I am interested in a tighter refiguring of these sites of activity for two reasons. The first is more general and is based on an argument that "service" is actually an epistemologically productive site of activity. It is this issue that serves as a framework for the chapter. My second reason for working toward a tighter configuration of research, teaching, and service comes specifically from the strengths, purposes, and applications of technical and professional writing. This discipline, perhaps more so than others, is immediately relevant to communities around a given university, is a powerful place from which to serve those communities, and is a discipline that will grow in sophistication from work outside the university.[1] What I have described in these last few sentences is not "mere" service but also combines teaching, program design, and research into a matrix of interests and activities. My argument is this: An approach to technical and professional writing that works toward a rearticulation of research, teaching, and service is a powerful way to do academic work and can positively alter the meaning and value of technical and professional writing itself as a site of activity.

TECHNICAL AND PROFESSIONAL WRITING, SERVICE LEARNING, AND PROGRAM DESIGN

If my experiences at conferences such as the Conference on College Composition and Communication, the Association for Teachers of Technical Writing, and the Council for Programs in Technical and Scientific Communication are any indication, service learning is increasingly common.[2] But why? In a sense, service-learning projects are an extension of technical writing pedagogies that have been in place for some time. The use of cases in writing courses, for example, is commonplace despite the feeling of some that the fictive scenarios provide inadequate audience constructs (for example, Artemeva, Logie, and St-Martin 1999). For many, including those whom I worked with at Georgia State University, cases provide a rich context for learning about writing, organizations, and other complex relations associated with writing (such as politics and ethics), and so we have written cases for a number of writing courses. Technical writing teachers have long used writing projects in which students work on solving problems for real clients, or what Huckin (1997) calls "community writing projects" (see the first few pages of Huckin's article for a sense of the number of programs that employed such pedagogical practices in the late 1990s). Yet, at the time Huckin wrote his article on technical writing and community service, he knew of no technical writing programs that employed service learning. That situation has certainly changed, and there are two reasons for this, I think: service learning has caught fire across the university within the last ten years, and technical and professional writing programs have been well positioned to embrace and enhance the pedagogy. Because of the focus on complex problems and real clients, then, service learning is in many ways a natural extension of pedagogies common in technical and professional writing classes.

My concern in this chapter isn't primarily with service learning, but rather with programmatic connections to service learning. Still, service-learning teaching is at the core of the changed practices I'm arguing for here, and so I begin with my approach to service learning. Like Huckin (1997), who articulated his goals for service learning in technical writing as (1) helping students develop writing skills, (2) helping students develop civic awareness, and (3) helping the larger community by helping area nonprofits, my goals for service learning are to take part in long-term community change by meeting the needs of community partners and to provide rich and compelling contexts for student learning. These goals are actually quite complicated in how they play out. In fact, they bleed into all aspects of our writing program and my work at Georgia State.

Setting up service-learning projects takes some time. The ultimate goal is to make service-learning programmatic (more on this later), but currently I am the only faculty member who consistently teaches courses with service-learning components. The process actually runs throughout the year. I have contacts at my university's office of community service learning who occasionally funnel projects my way. I am sometimes asked by our AmeriCorp program to speak at training and information sessions with members of community-based nonprofits. These opportunities often result in new projects and relationships. And I have created a network of contacts in Atlanta, with whom I have been working for nearly four years now. These efforts are essential because through them I am trying to build long-term relationships with organizations in the community that make a difference in people's lives; likewise, I am trying to make our professional and technical writing program an organization that also makes a difference in people's lives.

Depending on my teaching and that of interested colleagues, I work with my contacts to come up with seven to fourteen projects each semester, which meet the following criteria:

- The projects meet a real need as articulated by our community partner
- The projects are sophisticated and writing related[3]
- The projects fit into the course time frame (about ten weeks)

The heart of the criteria is that these projects must be of service to the people with whom we are working. When these criteria are met, I begin to address other constraints. Once potential projects are identified, I visit my contacts at their locations to learn about the organization, make sure the site and neighborhood are safe and accessible to students, and further discuss the contours of the project. If my contact person expresses the desire to proceed, I write a letter of understanding, which is based on my university's standard intern contract. This letter is then rewritten by the contact person, if need be, and eventually is given to students as well (who can also add to the letter). Finally, the contact person is invited to class for the first day of the project to see the university, to see the space where students work and learn, and to meet with the group of students who will be working with them.

The students who participate in these projects are diverse in terms of age, race, class, and gender—"typical" students at Georgia State. Although the course serves students from across the university, the majority of students are juniors and seniors from the business school. Technical writing, on the other hand, isn't a true service course. About half the students in technical writing enroll to fulfill requirements for majors other than English; the other half plan to be technical writers upon graduation. The diversity of student experience is useful for service-learning projects because not only do these students have a range of interests and expertise but many also work as professionals and so bring rich histories and skills to our classes.

My approach to service learning is somewhat different from the model typically presented in composition and some technical and professional writing forums. The difference is not really in the pragmatics of setting up or teaching a service-learning project; rather, it is in the institutional framework I am trying to create. Therefore, I am more interested in relationships between the writing program and community-based organizations than I am in student–community agency relationships. In composition studies in particular, the common service-learning model is to have students find projects to work on or to choose from a wide array of projects—usually more projects than can be addressed in a given semester. Student choice, student agency, student voice are valued and for good reason (see Bacon 1997).[4] One goal of service learning is to educate for citizenship and to transform students' understandings of and relationships with the world around them. But my concern with such an approach is that it too often sounds like a low-level colonization of the communities around a university.[5] I question, in other words, both the pragmatics and ethics of such an approach. I wonder if students can find appropriate projects with community organizations within the time frame allotted. But more importantly, my experience with nonprofits suggests that students continually coming to them looking for projects takes valuable time and tends to raise expectations that might not be met. These expectations can create situations that hurt rather than help university-community relationships. In other words, I have serious doubts about that ability of service learning to accomplish either its service or its learning goals without a solid institutional home. Communities can indeed be hurt if they are in fact *burdened* by responding to numerous students looking for projects, particularly if there is no solid commitment that the project will be completed. The primary goal, after all, is to help community-based organizations help their communities; it is to participate in community change. Yet, when the primary motivation and concern is student agency, student learning, and student growth, I think service learning runs a serious risk of doing harm.

To avoid harm (at the very least), service learning in technical and professional writing needs to be part of the writing program. And so let me outline here the theory of institutional design necessary for creating that home. First, meaningful service is connected to long-term and community-driven attempts at change. Students, by the very nature of their position, cannot make the long-term commitments necessary to participate in meaningful community change. Faculty and programs can and so should make themselves available to help communities; we shouldn't be sending students into communities, like missionaries, to find problems. Second, ongoing processes of community and institution building are integral to community change (see Kretzmann and McKnight 1993 for much more on community building and change). Writing teachers and students can participate in community building and change, but only to the extent that we move away from an individual service ethic, which I tend to equate with academic charity (individual classes and students serving others, for example), and toward a community-situated ethic that seeks sustainable change, which I tend to link to (community-defined) issues of justice. Writing programs are far more useful to communities than to individual students and faculty because they provide a context for meaningful student and faculty work. They can do so, however, only if they are designed with a community interface.

I am primarily interested is the relationship between community-based organizations and the writing program because such an institutional relationship is more powerful and potentially more transformative. Service learning must primarily benefit the community partners with whom we work—they must be given preference—and the best way to ensure that this preference happens is to develop meaningful, long-term relationships with them. These relationships must be institutional to be effective. They cannot depend on the charisma of individual students or the commitments of individual faculty members, although they almost always start that way.

At Georgia State, we are attempting to institutionalize service learning within the writing program. At the undergraduate level, we situate service-learning experiences in both technical and business writing classes. Although service learning has never been limited to these classes, they are concentrated within these curricular slots to ensure that students have the option of service-learning experiences during their time with us, an option we encourage. In a larger sense, then, relationships between the writing program and the community are part of the identity of the writing program. Such relationships affect not only what happens in the classroom but also the kinds of experiences we offer students, the types of classes we offer (and will offer), and ultimately, the work we do as a faculty and a program. Service and community involvement, then, flow into other categories of work, and each site of activity—service, teaching, and research—is potentially transformed. In the next section, I will use two service-learning projects to demonstrate this possibility.

REFIGURING RESEARCH, TEACHING, AND SERVICE

The two cases I discuss in this section began with relationships connected to my service-learning efforts. Each case shows how "service" activities can have intellectual substance; how "teaching" can both serve and foster research; and how "research" can serve and instruct.

The first case concerns my involvement with rethinking public policy efforts associated with the local Ryan White Planning Council (see Grabill 2000 for a more complete discussion). The Planning Council is a federally created body that makes decisions with respect to HIV/AIDS care in Atlanta. Most urban areas have such councils. The Planning Council must be composed of individuals who fit a number of categories (everything from health care providers to government officials), and at least 25 percent of the local Planning Council must be made up of individuals affected by the disease. In addition, the composition of that 25 percent must match the current demographics of the disease (which has become increasingly low-income, non-white females). The theory here is that those most affected by the disease ought to have a significant say in making policy about their care. However, meaningful client involvement isn't easy. In fact, the feeling of many involved with the Atlanta Planning Council is that meaningful client involvement hasn't been achieved: the council hears from too few clients, who represent a rather narrow range of those affected by the disease.

I became involved with the project to address problems of client involvement through a student's service-learning project in one of my technical writing classes. The project in question was completed with Kuhrram (Ko) Hassan, an adolescent-HIV/AIDS educator, who worked for one of the service providers funded by Ryan White legislation. Ko was concerned with generating and documenting client involvement at his agency, which became the focus of the student project. For part of a semester, the student worked with Ko to understand his position, the policies and procedures of the organization, and ways in which he and others at the organization interacted with clients. The student's goal was to create with Ko and others a process by which involvement with clients could be easily facilitated, recorded, and then written about and shared with others.[6] She (the student) produced a short procedures "manual" (a process-flow chart, really), some job descriptions relevant to this process, and a formal report documenting her research and arguing for her work (a "product" and a report are typical of the deliverables for projects like this).

The student project was complicated, and in many respects, the student never finished it (although she did well within the context of the course). During the course of the student project, however, Ko and I began to discuss the larger problem of client involvement that was affecting the Planning Council's policy functions. Our conversations eventually evolved into a research project with two interconnected goals: (1) to improve client involvement in policy making by creating with clients procedures that overcame current barriers and (2) to create documentation of client involvement for use in policy discussions and reports of compliance to the government. So I was invited to help address a problem, and this invitation was framed as research, which was important for the

Planning Council because it gave credibility to voices too easily dismissed as isolated, to evidence from clients too easily ignored as anecdotal, and to client concerns too often dismissed as complaints. For obvious reasons, framing my involvement as research was important to me as well. The time I devoted to this project was significant, and to frame it as "service" or even "teaching" within an institution that still maintains a hierarchy of research, teaching, and service was unwise. More to the point, however, the work I did with the Planning Council *was* research. But it was also a service to that organization and to the people with whom I worked, and initially it was an explicit part of my teaching.

The Planning Council project is important for other reasons as well. Because the project was one of my first service-learning experiences, what I learned changed how I teach technical and professional writing classes. I began to look for technical and professional writing practices in community-based contexts. I began to more fully understand the role of writing and research in public policy processes. And I began to rethink the common ways in which technical and professional writing identified itself as a discipline. For example, I have started to think about what might happen to technical and professional writing if we fully embraced civically focused, nonacademic writing and writing in noncorporate and governmental organizations as a *critical* concern. Certainly, the kinds of questions we would seek to answer in response to these different contexts would change. We would also teach different sorts of writing to a new group of students and collaborate with units within the university we don't currently work with—public health, city planning, and public policy programs, for example. Our writing program designs would similarly change. Service outside the university has been fertile intellectual ground for me because it has forced me to rethink the identity and social value of technical and professional writing both at Georgia State and within the discipline at large.

The second case concerns work that is ongoing. As part of the regular conversations I have with community-based organizations, I became involved with the United Way, who wanted to list our writing program as a "technical support resource" for their grant programs. Through these programs, the United Way funds grassroots organizations that form to solve specific problems in neighborhoods throughout Atlanta. We have worked with a couple of organizations funded by United Way grant programs.

In 1998, I was appointed to an ad hoc United Way committee investigating the use of Geographic Information Systems (GIS) technologies to provide data-based maps of neighborhoods and communities to be used by the United Way and other organizations for decision-making regarding needs and services. Students were to provide research and writing expertise to this project. Although that project was soon shelved, I developed a working relationship with Patrick Burke of The Atlanta Project's Office of Data and Policy Analysis, the group supplying the GIS and planning expertise.[7] We agreed that some kind of Web-based interactive database would indeed be a useful tool to neighborhood and other community-based organizations. Such a tool could help them participate more effectively in planning decision- making processes that demanded information and analysis that was tough to acquire (see Sawicki and Craig 1996 for one version of the theory driving this effort).

Work on this project still continues. Our goal is to design with stakeholders from community and neighborhood organizations a Web-based queryable database that returns data in the form of maps.[8] The tough part, of course, is designing it in such a way that it is usable by people with varying experience and literacy levels. In addition, our initial feedback suggests that this tool will be even more useful if it serves more communicative functions—if, for example, it contains spaces for exchanging ideas and spaces for matching people with people and people with resources. For me, this database is a major research project. But it has also become a regular part of my teaching and, through my teaching, a service to some communities.

Early in the project, when Patrick Burke and I were just exploring possibilities, a group of students helped me with research related to access to computers in the city of Atlanta. Our focus was on libraries and other centers that allowed free public access to computers and computer networks. For a Web-based database to work for people associated with grassroots and community organizations, there must be infrastructural access and a host of literacy-related accesses. The student project was focused on mapping infrastructural access in certain neighborhoods and gauging the literacy support (such as documentation) that would be necessary to use Web-based tools. What the students discovered was depressing at best, and it was an eye- opening experience for them, both in terms of the uneven distribution of information technology and in terms of how such inequality deeply affects the project we were working on. Students began to understand that although it would have been easy to design these Web-based tools in a vacuum, such a resource would have failed.

Later, once we had developed a prototype of the Web site, another group of students took on the task of designing the usability research necessary to write the online help, and one of these students, during a later independent study course, took on the task of writing an initial version of the help system. Like the first group of students, this group was forced to confront the complexity of a "real" project; their experiences were both frustrating and exciting. The richness of these contexts for instructional purposes cannot be overstated. The teaching benefits are substantial because they place students in complicated writing contexts and ask them to deploy many of their intellectual skills as writers toward developing a solution. Students also get an opportunity to work closely with faculty on research. This work is not only a relatively rare opportunity but also an intellectually meaningful one. Students don't often get to see what we do, and this observation strikes me as an important part of their education. At the same time, students are given an opportunity to understand aspects of their community and participate in community service. I highlight this project because, like the public policy project, it is an example of how service engagements are productive; and furthermore, it is an example of how students can participate in faculty research, thereby enhancing instruction.

I hope these projects have provided a deeper understanding of the transformative possibilities of service learning, but, more importantly, I hope they illustrate how a tight integration of research, teaching, and service begins to blur distinctions among these three areas, infuses each site of activity with shared energy and actions, and makes the work of students and faculty in technical and professional writing more meaningful and, in my mind, potentially one of the most radical sites of activity in the university.

IMPLICATIONS AND NEW DIRECTIONS

My suggestions for rethinking research, teaching, and service begin pragmatically with service learning. The process of service-learning design that I urge others to consider looks something like this model:

1. Develop relationships with others inside and outside the university who are supportive of service learning, who want to participate in service learning, and who are willing to assist in the development of service learning in technical and professional writing.
2. Work to integrate service-learning experiences into the writing program.
3. Begin slowly and small with a few service-learning partners and one

writing class; make sure beforehand that the projects serve the needs of the partner, are curricularly appropriate, and can be completed within a term's timeline.

4. Frame the service-learning projects appropriately in the classroom by showing how such work is meaningful for writing instruction and how service-learning experiences are one way to achieve the larger goals of the university. (Students, in my experience, need to be persuaded to take risks. Persuade them and then reward them for doing so.)
5. Follow up with students and service-learning partners to honestly gauge the student's intellectual and service experiences (reflection) and the nature of the benefit to the partnering agency or organization.
6. Maintain relationships.

This process is one way to institutionalize service learning and increase the possibility that such experiences transform the work of the university. Over time, what counts as "curricularly appropriate" will likely change as views of writing change based on experiences with community organizations. In addition, research opportunities will abound. I see more interesting projects each year than I can possibly do. These opportunities can also filter into graduate programs and theses and dissertations. My point, again, is that embracing and then institutionalizing community involvement is one way to transform the work of university faculty in substantial ways. Research, teaching, and service, as I hope I've shown, tend to blend into one another to more fully become sites of activity where one can teach, serve, and research at the same time.

There are certainly limitations to the approach I have presented. It is time consuming, and project selection and management can be difficult. In addition, when projects don't work well, it causes significant problems for students and community partners. And because the service elements focus on organizations (when we work with organizations that serve the homeless, for example, students may not work directly with the homeless), students don't always see their work as service and don't always reflect on their work in ways that are meaningful. In terms of the larger argument I have been making, it is also telling that I still feel the need to validate the time spent and the work accomplished in terms of my ability to generate research from community-based projects. Finally, my insistence that the more transformative aspects of refiguring research, teaching, and service depend on institutionalizing service learning places a burden on writing programs that some may see as too significant. It remains to be seen whether we can handle the burdens here at Georgia State.

I see these problems as program design specifications, and I think the following specifications, which move beyond the more concrete service-learning suggestions at the beginning of this section, are more generally useful design heuristics:

- Service learning is unsustainable within and outside the university as an individual initiative; it does not respond well to community-driven needs and will exhaust individual faculty. Service learning must therefore be part of the writing program's design and therefore integrated into the curriculum (but not just slotted into classes; it can and should transform how classes are taught), given administrative support, and rewarded as part of faculty work, perhaps much like the single course release given to most internship supervisors.
- Service learning must be seen as substantive and intellectually rich work and so must be visible and presented in these terms. Its meaningful presence in the writing program and curriculum is one type of visibility, as is its ability to generate research. Service must also be visible in other ways as well, such as a university Web presence in which those involved with service learning can describe its value to the community (through the use of project evaluations, letters of thanks, and project artifacts, for example). Like all good programs, technical and professional writing programs can and should argue for their social value. Service-oriented writing programs help.

Community based work in technical and professional writing allows technical and professional writing students and faculty to work across a number of contexts, with diverse audiences, and on projects of civic significance. I see technical and professional writing as a site of truly radical activity because of our ability to redefine the work of university faculty and students, because of our ability to move among the university, the corporation, and the community, and because of our ability to understand the powerful ways in which writing constructs institutional systems and changes them. We do work of "substantive idea content and significance," and we do it across sites of activity long artificially separated. Continuing this work should be at the center of who we are as teachers, researchers, and citizens.

6

NOTES TOWARD A "REFLECTIVE INSTRUMENTALISM"

A Collaborative Look at Curricular Revision in Clemson University's MAPC Program

Kathleen Yancey
Sean Williams
Barbara Heifferon
Susan Hilligoss
Tharon Howard
Martin Jacobi
Art Young
Mark Charney
Chris Boese
Beth Daniell
Carl Lovitt
Bernadette Longo

The faculty of Clemson University's MAPC program—rhetoricians and professional communications specialists of various kinds—gather this Monday, as we do weekly, to continue our work: designing, implementing, and enhancing the "MAPC," our M.A. program in professional communication. Rich in theory and practice, it's a program benefiting from the attention provided by frequent faculty discussions. Our current task is one that continuing graduate programs take up periodically: reviewing and revising a reading list for a comprehensive exam, in our case the MAPC oral exam keyed to the MAPC reading list.

Today, we begin our discussions by focusing on a central programmatic issue for all technical communication programs, raised in an email sent by a new colleague, Sean Williams:

> I had two students in my office this week trying to figure out just what on earth social construction has to do with writing a memo and why they need to know Cicero to write a good proposal. "Just give me the format, Dr. Williams, and I'll write it," they say in not so many words. I think this is a huge curriculum issue, too, at the grad level because the perceived bifurcation (is that word too strong?) of the program begs the question of "fundamental" knowledge for proceeding in the program. Why aren't students required to take 490/690, "Technical Writing," but are required to take classical rhetoric? I don't mean to imply that they should be separated because I don't think they should be.

> However, I'm not sure that we as a faculty are clear on exactly how the areas are connected, and the result is confused students and perhaps a confused faculty. We need, IMHO, to articulate, in writing, goals that unite the two threads in a mission statement or something like it because this type of focused attention on "What do we do?" necessarily precedes "How do we do it?" Revising the reading list is a "How do we do it?" consideration. And, not to be too self-aware, but would defining "what do we *do?*" be reflective instrumentalism?

In his recent *Collision Course: Conflict, Negotiation, and Learning in a College Classroom,* Russel Durst (1999) tracks the competing agendas of students and faculty in first-year composition studies classrooms. Like our MAPC students, Durst's composition students want practical help; like Durst's colleagues, we faculty want theory and critique as well. It's another version of the theory/practice divide, with faculty on one side, students on another, what Durst—and before him, Patrick Moore (1996, 1999), Carolyn Miller (1979, 1996) and Robert Johnson (1998a, 1999), among others (see Bridgeford 2002)—couches as a conflict between two impulses: on the one hand, students' "instrumentalism" and on the other, faculty theorizing.

Durst's (1999) curricular reply to this tension is what he calls "reflective instrumentalism," which, he says, "preserves the intellectual rigor and social analysis of current pedagogies without rejecting the pragmatism of most . . . students. Instead, the approach accepts students' pragmatic goals, offers to help them achieve their goals, but adds a reflective dimension that, while itself useful in the work world, also helps students place their individual aspirations in the larger context necessary for critical analysis" (178).

Which leads us to ask the following questions:

Would Durst's concept of reflective instrumentalism provide a useful way of framing our program in professional communication? If so, what changes to the program might it recommend?

Would other concepts already part of the culture of the program—such as "professionalism" or "reflective practice"—provide framings more congruent with the program? What changes might they recommend, particularly if they were made a more explicit or integral part of the program?

Are there other ways we might think about the program, especially about the relationship we seek to establish between rhetoric and technology?

How might we use these framings to develop a language to explain our expectations to students—and to ourselves?

In the pages that follow, we'll take up these questions as we narrate the process of revising the reading list for the MAPC. We'll approach this task as participant-observers of our program and our processes of curricular design. Additionally, in narrating the processes that we used in our curricular decisions, we'll explore the possibilities for representing these processes textually and our rationales for why we choose to represent them as we do. Our reading list is, of course, only one of many representations we could make of our process: other representations include MAPC recruiting materials, our MAPC handbook, and MAPC graduates themselves.

In conducting this study, then, we hoped to build an understanding of

1. the processes we used to review our program
2. the ways we represent that process textually in different rhetorical situations
3. a consideration of what those representations do to the process and our understanding of it

In other words, we want to consider a final reflective question: what does the means of representation suggest about the program itself, and how will it affect the very program under scrutiny?

Equally important, we hope that, in creating this reflective account of our revision process, we make a successful argument that other programs might also try such a collaborative revision themselves. Such curricular revision isn't often consciously observed or reported on, nor is it often theorized, yet (ironically) given its influence on students, it's critical. The key factor, as we found, is to *work together.* In other words, we chose not to assign this task to a subgroup of a larger committee or to a special task force, but to take it up as a committee of the whole. We knew in proceeding this way that the process would take more time, would be more cumbersome, would require considerable negotiating skills. We understood that, vested as we all are in what we think is important, we were taking a risk, that negotiations could break down, even fail. At the same time, we found, and we think others will as well, that both process (articulating together our goals for the program and ways these are realized in a set of readings) and product (the revised list) are worth the risk and effort.

We have many ways to narrate the story of our process, all of which comment differently on the values of the program itself. We could simply record it, for instance, by noting that we began work on the reading list in the fall and concluded in the spring. We met weekly, some of us routinely, others as time and other responsibilities permitted—as we taught classes, wrote papers, attended conferences, recruited new faculty, developed a new undergraduate Writing and Publication Studies major. Representing the process this way indicates that *major curricular issues are a matter of course for the committee of the whole,* responsibilities that we took up seriously. That itself is both claim and statement about the program.

We could also tell our story through numbers. We began with twelve categories, including among them topics that were identical with our five core courses: Visual Rhetoric; Workplace Communication; Classical Rhetoric; Introduction to Professional Writing; and Research Methods in Professional Communication. Included as well were other categories that seemed to play a role in the program, although the role wasn't always clear: Literacy, Technology, International Communication. We began with forty-four items and were committed to maintaining that number, to resisting the impulse to grow the list. To accommodate the impulse, yet stay close to our target number of items, we created an archival list of all the items that could be included and worked from those. We also spoke as though all the categories were equal, although as individuals we had preferences, and it wasn't difficult to discern what those were. Given the number of items and the number of categories, each category—from Classical Rhetoric to International Communication—seemed eligible for about four entries. Representing the process this way would indicate that we value a certain conservative structure, that we like to explore the possibility of expanding our reach, but that at the end of the day we like to come home to the familiar, where everything has its place.

We could also tell our story through understandings, specifically our understandings about the process we should use for revision. Some of us thought that we should use the old list as a point of departure and should proceed by revising this list, understanding that to add a new text, we had to drop an old one. Some of us thought we should work from a blank slate to build a new list. Some of us thought we should focus on certain underrepresented areas—technology and diversity among them. Some of us felt passionately about our favorite figures and texts and thought that others should see the list through our theoretical lenses. And when passions were strong, we used our communication symbol—a "Fight Club" button, a promotional pin from Brad Pitt's movie by the same name popular at the time—to signal that an individual had become overly invested in their personal preferences. The "Fight" button—which even now is seen by some as sign of negotiation, by others as sign of friction—became a part of the process, a material token of the work to which we are all committed.

• • •

We are choosing, however, to tell our story *primarily* through our individual voices, in part because this individualism is ultimately what we value in the MAPC program and hope to teach students: a respect for a multiplicity of voices, perspectives, personalities, and passions. In part, we hope, through this way of telling the story, to work in *palimpsest* (de Certeau 1984; Barton and Barton 1993), to include in our collective story here traces and vestiges of how it came to be. In other words, the new reading list itself is one map to the program. But how that map was created can itself be mapped, and that too is our aim.

Our listserv makes such a representation possible. Listen in as we enter *in medias res* to Mark Charney, the chair of MAPC, summarizing the review of one meeting:

> Dear Kathleen (and MAPC Committee): Here are the best notes I can muster up from the meeting you missed. Please forgive me if I've misrepresented anyone! We discussed primarily visual communication, and plan next week to discuss professional communication theory, ideology, and teaching/pedagogy, so please, MAPCers, come to the meeting with good notes about what you want to do with each of the next three fields.
>
> Sean began the meeting not only with great ideas about vis comm, but also samples and examples of each of the following.
>
> 1. Jacqueline Glasgow's "Teaching Visual Literacy for the 21st century" for its emphasis on decoding images, making passive observers active, and its explanation of semiotics.
> 2. Williams and Harkus's "Editing Visual Media" for its emphasis on the verbal vs. the visual, especially its practical bent (and the good example of a ball vs. a basketball, etc.)
> 3. A PRIMER OF VISUAL LITERACY, Chapters Two and Three: one offers guidelines for visual literacy, a good overview, and the other, basic elements of visual communication.
> 4. Edward R. Tufte's new chapter 2 in VISUAL EXPLANATIONS, and chapters 4 and 5 in the old book to keep terms like Chat Junk, etc.

5. DESIGNING VISUAL LANGUAGE by Kostelnick and Roberts, especially chapters 1 and 2 which tie rhetoric to visual.

Now here is where I break down in terms of who suggested what. Both Tharon and Chris had a say here, and all three agreed, as did the rest of the committee there, about the worth of the [texts] below. It wasn't a fight club situation at all, and we got through this in record time, so much so that it surprised us into being unprepared to move on, so we adjourned early! (well, ok, only a few minutes early)

6. Kress and Van Leduwen's READING IMAGES, Chapters 2 and 3 about narrative theory and visual communication, especially the linguistics of visual design. Also, possibly chapter 4 which deals with modality.
7. Karen Schriver's DYNAMICS IN DOCUMENT DESIGN, pages 168–181 (this was Tharon's I remember), which deals somewhat with usability studies and technical brochures.
8. Carl mentioned a new book by Kenneth Hager with one chapter on Visual Communication. He plans to give us the exact reference next week.
9. Sean encouraged us to keep the Elizabeth Keyes already on the list, while everyone finally agreed to keep everyone already there, especially Barton and Barton, who everyone agreed was a clear introduction for uninitiated students, and Maitra/Goswami, the Kostelnick on the list, etc.
10. Also, EDITING: THE DESIGN OF RHETORIC, the final chapter about typesetting and production, was mentioned as something that may help basic students. By Sam DRAGGA and GWENDOLYN GONG.

Some discussion ensued about how this was often the first class most students took and how it has to begin very basically. The Hilligoss book was mentioned by Barbara Heifferon, who uses it successfully in the classroom, but using it would break our rule not to use our own texts in the classroom.

Finally, we agreed to make two lists—one the reading list for orals, and the other, a list of all of these related texts, each significant to students in the field and to students researching theses and projects. Such a list could be updated every year for the orientation MAPC book, making that list a current one from which we could update the orals list anytime we wanted.

I apologize if this is rough, or if I've given credit to anyone for something he or she may not want! See you next, and every Monday . . .

Mark Charney

Sometimes discussions that seemed to be about one issue—the one previously mentioned about visual rhetoric, for instance—turned out to be about others; and always in the background was the question: *Who is the MAPC student?*

> It would be almost impossible to define the ideal student.
>
> Beth Daniell

> I agree. What I'm driving at in using this term is actually something like "what should every MAPC student leave the program with?" I'm not thinking here in terms of discursively forming ideal students, but rather of a minimum set of qualifications and knowledge that all students should possess, much like the list that you offered: theory, practice and technical expertise. The tricky part is figuring out what "theory," what "practice," and what "expertise" we're talking about. Is theory rhetorical theory or is it professional communication theory? They're related certainly, but not by any means the same. Is practice, writing seminar papers or creating multimedia? Again, they're related, but not the same. Is expertise a theoretical expertise or knowing how to use computers well?
>
> The separations are a matter of emphasis and it seems to me that this emphasis needs to be fleshed out a little more by having conversations like that we had today. It was EXTREMELY helpful in helping me to understand the way the people in this program view what the program does. Now that I have a little more context on "what it is that we do" I can make more informed choices about what to include/exclude from the reading list.
>
> Sean Williams

Who is the MAPC student? This question haunted the process, as we understood our role in defining and constructing that figure. Still the student, as Beth Daniell suggests, eludes us.

> Continuing to beat this poor horse, I don't think we can always be more specific. I want students to have some sense of technology. What does that mean? I don't think it means everyone has to design Web sites. I think a lot of it is what the student wants it to mean. They are agents in this process, not empty containers. While I understand your need for definition, I have been teaching way too long to think that my categories or yours are adequate to cover all the students. We set up the framework in which individuals and teachers work. The outcome is not up to us. I'm constantly amazed at what my students come up with—and like you, they often complain that I am not being clear enough.
>
> Beth Daniell

At some point, our negotiations on *who is the MAPC student* turned from abstract to particularized as we began to horse-trade—"I'll trade you a Landow for a Plato"—to represent what we thought every MAPC student should know. To accomplish this, we all forwarded nominations for each category, not to select winners, but to show patterns. We called them *tallies.* (Language matters.) It wasn't a flawless process, and it provided a set of questions that continue to beguile us.

How to negotiate?

As implied previously, having a written record helps; here Kathleen Yancey provides context for understanding the tallies.

Draft of Nominations for Reading List for MAPC

Context: Not everyone sent in tallies. Not everyone voted for three per category, so I just counted the number up to three. Not everyone sent only three per category, so I just counted the first three. If you numbered them, I took the top three.
A couple of suggestions appeared that had not been mentioned or discussed previously. I did not include them on the list.

Issues: Presence?
Absence?
The categories: do we need all of them? Two folks mentioned that they would dispense with teaching, one that we could dispense with literacies.
Should all categories be equally weighted?
What's the role of the current list?
How do these items compare to what's on the current list?
Some items are repeated—Faigley and Barton and Barton—come to mind.
Can we cross-reference some items?
Is diversity sufficiently represented?
Is technology sufficiently represented?
When we look at the list, what student have we constructed?

In the background, as we sorted through the tallies, discussions related to our questions continued. A major discussion involved the relationship of rhetoric to professional communication, as Martin Jacobi explains:

> I guess I'm wondering still what constitutes "rhetoric" for you. I'm hoping it's not something like "bombastic discourse having no relation to the real world,

to what professional writers—whoever they may be—do for a living." I'm trying to imagine the nature of "professional documents outside the frame of rhetoric" but I'm coming up empty. When Ornatowski talks about the engineer who has to write a report that will sell to potential customers an engine that will not start in cold weather, he talks about the rhetorical choices—and ethical choices (since any action, as opposed to motion, is necessarily ethical invested)—that the engineer is making. It's clearly a rhetorical document that Ornatowski's engineer is talking about.

I would agree that reading Aristotle is not the most effective way of teaching or learning ethics, but what's your point? If you're saying that pro com uses case studies and not theory to do things, then aren't you contradicting your earlier claim that pro com is theoretically sophisticated? Aristotle pointed out that he wrote his Art o' Rhetoric because teachers of rhetoric were only using something like case studies for their students.

Sometimes in the middle of all this discussion would arrive a listserv post from somebody outside our dialogue that reminded us that we were hardly alone in sorting out these issues:

> Last spring in *Time* or *Newsweek* there was a big article on Careers[,] and Technical Writing was featured heavily as a good bet for college students. We used it to help us bolster our argument for an interdisciplinary graduate certificate in professional writing.
>
> Irene Ward, from WPA-L (listserv)

The horse trading continued. It was smart; it was social; it was (of course) rhetorical. We made connections between other professional contexts and this one; we used such comparisons to think about what would best help students.

I'm thinking ahead to our next meeting and urging everyone come with a text or two to "be flexible" on. I think we are good enough horse traders to do this? Our task is not as daunting as it may seem. I counted 54 texts, and if we get down to 45 (shoot for less and see how that goes), that's only 9 to give up. I came up with that number because we are doing fewer chapters in Latour and Woolgar (e.g.) and others.

I'm also reading for absences. As peer reviewer for *TCQ* . . . I've reviewed a number of tech comm pieces. . . . My reviews have included some alarm about lack of awareness of something other than our good ol Yanqui point of view (I realize how strange this sounds in S[outh] C[arolina]). When I get the reviews back with other reviewers' comments as well, they are picking up on the same thing. All this to say that I'm concerned that we may not yet represent a voice of someone other than ourselves for the good of our students who will go out and work in a world that, surprise, does not look exactly like us. Thus we need at least Freire on board or someone that makes this point. There may be a better rep. I'm open. Unless I missed something on the list, I don't see us doing this.

I wouldn't mind trading a Doheny-Farina and Harraway for a chapter from Harding that addresses a couple of absences. The one that covers standpoint theory (also one of Tharon's lenses in his book) might serve. It's at home, I'll send the chapter # later. Harraway is so dense, though God(dess) knows I love her, she makes Vitanza read like a Sunday school picnic (most likely an abominable mixed metaphor).

Barbara Heifferon

Trading itself, of course, isn't an easy process. We understood our choices as signs, as representations. We read multiple gestalts into such a list, as Chris Boese self-referentially suggests:

Chris won't give up Harraway. And Chris wants Freire. Classic struggle with canons. You know what it is. For new points of view to come in, something sacred has to go.

I'm not trying to be intransigent here, but I have a different point of view on the list. The old list is dangerously deficient in the area of technology. Quite a bit needs to go in there to bring it up to speed with what is going on in the world. I am as much of a horsetrader as anyone, but I don't think technology should be the thing that has to "give" as much as other areas do. Of course reasonable people may disagree. But if serious room for technological issues and technology criticism isn't made on the list, I believe there will be major credibility problems with it.

Other areas have long held place on the list. Like Rhetorical Theory. They are the 900 pound gorillas. Technology scholarship is newer and having great impact in the field, changing the landscape of the field even as we speak. If our list doesn't make room for it, it won't be because tech is a yearling gorilla, it will be because those of us who advocate for it haven't done a good enough job in making the case. The field is changing, with or without us. We just have to decide if we want our list to actually reflect that change.

In the end, as Barbara Heifferon's concluding post attests, the process worked—

I wanted to tell you the good news in case no one else had. . . . At MAPC today, after a meeting that lasted under an hour, we went from a reading list

> of 62 down to 46!! Trades were made and collegiality remained intact after a few vigorous conversations. . . . I think it's a great list. I took notes as did Mark, and someone will get the final list ready for fall!! We did cheat a bit (folding a few readings of same authors together, just a couple)

—if by "worked" we mean that we had a new list that most of us would agree was better and that we had negotiated well. The list: it's appended. It's not perfect. But *most* of us would agree, on *most* days, that it's better. And it's different: some eleven items are new. Some of our favorites—from Harraway to Bakhtin—didn't make it. But they are on the archival list, and they are available for another (negotiating) day. And although second-year graduate students have been given the transitional option of using the old or new list, the new students are using this list, and we are finding it a better fit for most of the core courses. As Sean Williams, Barbara Heifferon, and Kathleen Yancey (2000) put it at CPTSC 2000, "Students who have seen the new list make positive comments about it because the list manages to bring what seem to be opposite poles—reflection and instrumentalism—into a single reading list that represents the current state of our discipline." (See http://www.cptsc.org/conferences/conference2000/Williams.html.)

• • •

We began this chapter, as we began the revision of the list, with an interest in bringing theory and practice together. The new list doesn't completely resolve this divide because we ourselves are still resolving it; probably we should have understood that it's too large and too complex a divide for this single curricular practice to resolve. But we have seen that we can negotiate: we can compose a list that constructs a student we'd like to see develop within our program and whose development is fostered by our new reading list. The program, in other words, is dynamic: it is able to accommodate both change and the tensions accompanying such change.

As Bernadette Longo puts it,

> Now that we've gone through one iteration of this process for revising the MAPC reading list, it seems that we've played out the issue that motivated this revision in the first place: "the perceived bifurcation . . . of the program" between theory and practice. We entered this process on high theoretical ground, positing topics that should be included in a reading list that reflected the important conversations in our field. (Actually, I'm not sure we agree on what our field is, but that's another chapter.) We all put forward readings in

> these categories based on theories and philosophical points of view informing our own research and teaching. But as weeks went by and the discussions ground on, it seemed that we slipped unnoticed into the arena of practicalities as the size of the list and the pressures of compromise constrained us. By the end of the year, many of our discussions were shaped by the need to keep the list at about 44 items and also to include representative works from all 12 of our original topic areas. . . .
>
> The intent of revising the list was theoretical, but the revision process turns out to be mostly practical. Once again, questions of "how do we do it" seemed to overwhelm questions of "what do we do." As Sean has mentioned in postrevision discussions, I'm not sure we have a handle yet on the question of what we do (as a program) when we shape our MAPC students' graduate studies. I think we have come up with a more current reading list through this process, and that's good. I'm not convinced, though, that we have better articulated the intent and objectives of our graduate program. Maybe that discussion needs to take place separately from the reading list revision process.

Which, of course, it has, through our later discussions—on exams, on projects and theses, on discussions about the kinds of experiences we hope to offer students.

Ultimately, that we didn't resolve the theory/practice split, or that Durst's (1999) construct didn't inform the entire process, or that we all feel there are still some gaps in the list doesn't matter as much as it may appear: this is not an exercise in Katz's (1992) expediency. What ultimately matters is that in the processes of (1) renegotiating our reading list and (2) negotiating the way we have chosen to represent it here, we discovered that we can practice what we preach to students: that successful communication, even involving the creation of reading lists, requires recognition and negotiation among many competing voices. In Durstian terms, we have had it both ways: in instrumental terms, we both accomplished the task and continued to reflect on the list, on the program, on the processes informing it, and on ways to weave together theory and practice into a coherent curricular whole.

In thinking about how and why such a process might be useful for others, we'd observe

- that participating in such a curricular revision can be a significant socializing activity, certainly for new faculty members, but also for more senior faculty as they interact with their new colleagues and with the possibilities for curricular revision;

- that it provides all faculty with a chance to examine how the field—and even the definition of the field—has changed since the last list was constructed;
- that engaging all program faculty in developing and maintaining a graduate program seems to require the kind of commitment realized in curricular negotiations and that these negotiations may entail friction and require delicacy and humor;
- that after having participated in this process, faculty understand the rationale explaining why individual readings are on the list as well as how the readings relate to each other, and they therefore are more inclined to see the list as a total package (rather than a set of disparate readings) and can explain this to students;
- that a reading list is just that, only a list; in a healthy curriculum, any list is necessarily and always penultimate given its contextualization within many other readings and experiences and the fact that it too will be revised;
- that the value of the list is likewise never fully understood until it is used by students and faculty together; and
- that what we have outlined here—by specific observations and linguistic montage—is a process, one more difficult and less efficient than if we had tasked it to a smaller group, but one more rhetorically productive. We created an opportunity to bring people together to communicate about things that matter: to write the program representing us and constructing students.

In short, we modeled for students the ways we'd like them to behave. The best we could do for students is to maintain a vestige of this idea in the reading list—and we did. We think this, too, may be one of the benefits of a collaborative curricular design.

THE READING LIST

Aristotle. Excerpt from *On Rhetoric: A Theory of Civic Discourse.* Trans. George A. Kennedy. New York: Oxford U P, 1991. (Book 1, chapters 1–3 [25–51]; Book 2, chapters 18–26 [172–214]).

Barton, Ben F., and Marthalee S. Barton, "Ideology and the Map: Toward a Postmodern Visual Design Practice," *Professional Communication: The Social Perspective.* Eds. Nancy Roundy Blyler and Charlotte Thralls. Newbury Park: Sage, 1993. 49–78.

Bitzer. "The Rhetorical Situation." *Philosophy-and-Rhetoric* 1.1 (1968): 1–14.

Bizzell, Patricia. "Foundationalism and Anti-Foundationalism in Composition Studies." *PRE/TEXT* 7.1–2 (1986): 37–56.

Blyler, Nancy Roundy, and Charlotte Thralls. "The Social Perspective and Professional Communication: Diversity and Directions in Research." *Professional Communication: The Social Perspective.* Eds. Nancy Roundy Blyler and Charlotte Thralls. Newbury Park: Sage, 1993. 3–34.

Bolter, Jay David. Either Chapters 1, 2, and 13 from Writing Spaces, or Chapter 1 from *Remediation.*

Brandt, Deborah. "Accumulating Literacy: Writing and Learning to Write in the Twentieth Century." *College English* 57 (1995): 649–68.

Brown, Stuart. "Rhetoric, Ethical Codes, and the Revival of Ethos in Publications Management." *Publications Management: Essays for Professional Communicators.* Eds. O. Jane Allen and Lynn H. Deming. Amityville: Baywood Publishing, 1994. 189–200.

Burbules, Nicholas. "Rhetorics of the Web: Hyperreading and Critical Literacy." *In Page to Screen.* Ed. Illana Snyder.

Burke. Kenneth. "Terministic Screens." Language as Symbolic Action. Berkeley, CA: U of California P, 1966. 44–62. An Excerpt from *Grammar of Motives.*

Debs, Mary Beth. "Corporate Authority: Sponsoring Rhetorical Practice." *Writing in the Workplace: New Research Perspectives.* Ed. Rachel Spilka. Carbondale: Southern Illinois UP, 1993. 158–70.

Doheny-Farina, Stephen. "Confronting the Methodological and Ethical Problems of Research on Writing in Nonacademic Settings." *Writing in the Workplace: New Research Perspectives.* Ed. Rachel Spilka. Carbondale: Southern Illinois UP, 1993. 253–267.

Duin, Ann Hill. "Test Drive-Techniques for Evaluating the Usability of Documents." *Techniques for Technical Communicators.* Eds. C. Barnum and S. Carliner. NY: Macmillan, 1993. 306–35.

Faigley, Lester. "Nonacademic Writing: The Social Perspective." *Writing in NonAcademic Settings.* Eds. Odell, Lee, and Dixie Goswami. New York: Guilford P, 1985. 231–48. And Chapter 1 of *Fragments of Rationality.*

Flower, Linda, et al. "Revising Functional Documents: The Scenario Principle." *New Essays in Technical and Scientific Communication.* Eds. Anderson, Paul, R. John Brockman, and Carolyn Miller. Farmingdale, NY: Baywood, 1983. 41–58.

Foucault, Michel. "Order of Discourse." In Bizzell and Herzberg, *The Rhetorical Tradition.*

Freire, Paulo. "The Adult Literacy Process as Cultural Action for Freedom and education and Conscientizacao." *Harvard Educational Review* 68.4 (Winter 1998): 480–520.

Gorgias. "Encomium of Helen." *The Rhetorical Tradition: Readings from Classical Times to the Present.* Ed. Patricia Bizzell and Bruce Herzberg. Boston: Bedford Books of St. Martin's P, 1990. 40–42.

Herndl, Carl. "Teaching Discourse and Reproducing Culture: A Critique of Research and Pedagogy in Professional and Non-Academic Writing." *College Composition and Communication* 44 (1993): 349–63.

Johns, Lee Clark. "The File Cabinet Has a Sex Life: Insights of a Professional Writing Consultant." *Worlds of Writing: Teaching and Learning in the Discourse Communities of Work.* Ed. Matalene, Carolyn. New York: Random House, 1989. 153–87.

Kaplan, Nancy. "Ideology, Technology, and the Future of Writing Instruction." *Evolving Perspectives on Computers and Composition: Questions for the 1990s.* Ed. Gail Hawisher and Cynthia Selfe. Urbana, IL: NCTE, 1991. 11–42.

Katz, Steven. "The Ethics of Expediency: Classical Rhetoric, Technology, and the Holocaust." *College English* 54 (1992): 255–75.

Killingsworth, M. Jimmie, and Jacqueline Palmer. "The Environmental Impact Statement and the Rhetoric of Democracy."*Ecospeak: Rhetoric and Environmental Policies in America.* Carbondale: Southern Illinois UP, 1992. 162–91.

Kostelnick and Roberts. Chapters 1–2 of *Designing Visual Language: Strategies for Professional Communicators.* Allyn and Bacon, 1998. 3–78.

Kostelnick, Charles. "Cultural Adaptation and Information Design: Two Contrasting Views," *IEEE Transactions on Professional Communication* 38.4 (December, 1995): 182–96.

Kuhn. Preface and Intro to *The Structure of Scientific Revolutions.*

Kynell, Teresa. Chapter 6, "1941–1950: The Emergence of . . ." *Writing in a Milieu of Utility.* Ablex, 1996. 75–88.

Latour, Bruno, and Steve Woolgar. Chapters 1, 4 and 5 of *Laboratory Life: The Construction of Scientific Facts.* Princeton, NJ: Princeton P, 1986. 15–233.

Laurel, Brenda. Chapter 1 from *The Computer as Theatre.* Addison-Wesley, 1993.

MacNealy, Mary. *Strategies for Empirical Research in Writing.* NY: Allyn and Bacon, 1999.

Maitra, Kaushiki, and Dixie Goswami. "Responses of American Readers to Visual Aspects of a Mid-Sized Japanese Company's Annual Report: A Case Study." *IEEE Transactions on Professional Communication* 38.4 (December, 1995): 197–203.

Miller, Carolyn. "What's Practical about Technical Writing?" *Technical Writing: Theory and Practice.* Eds. Fearing, Bertie, and W. Keats Sparrow. New York: MLA, 1989. 15–26.

Perkins, Jane. "Communicating in a Global, Multicultural Corporation."

Plato. "Phaedrus." Trans. W. C. Helmbold and W. G. Rabinowitz. New York: Macmillan Publishing Company, 1956.

Myers, Greg. "Texts as Knowledge Claims: The Social Construction of Two Biology Articles." *Social Studies in Science* 15 (1985): 593–630.

Nielsen, Jacob. Chapter 2 "Page Design" In *Designing Web Usability.* Indianapolis: New Riders Publishing, 1999.

Redish, Janice, Robbin Battison, and Edward Gold. "Making Information Accessible to Readers." *Writing in Non-Academic Settings.* Eds., Lee Odell and Dixie Goswami. New York: Guilford P, 1985. 129–53.

Schmandt-Besserat. "The Earliest Precursor of Writing." *Scientific American* 238 (1978): 50–59.

Schriver, Karen. Excerpt from *Dynamics in Document Design Creating Texts for Readers.* New York: John Wiley and Sons, 1997. Pages 168–181.

Selfe, Cynthia L., and Richard Selfe. "The Politics of the Interface: Power and Its Exercise in Electronic Contact Zones." *College Composition and Communication* 45.4 (1994): 480–504.

Sullivan, Patricia. "Taking Control of the Page: Electronic Writing and Word Publishing." *Evolving Perspectives on Computers and Composition: Questions for the 1990s.* Ed. Gail Hawisher and Cynthia Selfe. Urbana, IL: NCTE, 1991. 43–64.

Tebeaux, Elizabeth. "Technical Writing in Seventeenth-Century England." *Journal of Technical Writing and Communication* 29.3, 1999. 209–54.

Tufte, Edward. Chapters 4 and 5 of *The Visual Display of Quantitative Information.* Cheshire, CN: Graphics P, 1983. 91–105, 107–21. And Chapter 2 of *Visual Explanations.* Cheshire, CN: Graphics P, 1997. 27–54.

Vitanza, Victor. "Historiography."

Winsor, Dorothy. "Engineering Writing/Writing Engineering." *College Composition and Communication* 41.1 (Feb 1990): 58–70.

Yates, JoAnne. Chapters 1–2 of *Control through Communication: The Rise of System in American Management.* Baltimore: John Hopkins UP, 1989. 1–64.

PART TWO

Pedagogical Practices

7

STORY TIME

Teaching Technical Communication as a Narrative Way of Knowing

Tracy Bridgeford

Telling stories is the basis of how I teach—not just technical communication but any subject—composition, editing, literature, and publications management. I don't tell these stories simply to entertain students or to keep them interested—although certainly stories can perform that function. I tell stories because stories are a part of the practices of everyday life; they make it possible to articulate these practices. We know each other, our communities, and the world through the stories we tell each other about what we know, how we know what we know, and why we know what we know. Specifically in my technical communication classes, I ask students to read a particular story as a context for assignments and discussions. This approach helps students to contextualize the constructs and implementation of knowledge demonstrated in technical documentation—audience analysis, invention, information design, and documentation. In this chapter, I describe this approach, providing some example assignments and student writing in order to demonstrate how stories help me realize my pedagogical goals.

This approach is heavily influenced by Michel de Certeau's (1984) concept of stories as the articulation of everyday practice and Jerome Bruner's (1981 and 1990) discussions about *hermeneutic composibility*—how stories are made. These two perspectives provide the foundation for the construction of technical documents from a narrative perspective.

First, telling stories, or what Michel de Certeau calls the "narrativizing of practices," is a "textual way of operating" or "way of thinking" that involves a meshing of what one knows (theory), how one knows what one knows (practice), and how one applies that knowledge to situations *(metis)*. The telling of stories is characteristically concerned with the "style of tactics" (79), or a way of operating that traverses schemas into opportunities for action. Because tactics are opportunistic, they belong, de Certeau says, to the classical concept of *metis:* a "form of intelligence that is always 'immersed in practice,' which combines 'flair, sagacity, foresight, intellectual flexibility, deception, resourcefulness, vigilant watchfulness, a sense of opportunities, diverse sorts of cleverness, and a great deal of acquired experience'" (81). To be effective, stories must demonstrate this level of cleverness in their realization.

Stories are realized in the act of telling. Because a story "makes a hit (a *coup*) far more than it describes one," "its discourse is characterized more by a way of exercising itself than by the thing it indicates" (79). In this way, narrativizing is an "*art* of saying," or a "know-how-to-say," "characterized more by a way of exercising itself than by the thing it indicates" (de Certeau 1984, 78, 79): it is an "art of speaking . . . which exercises precisely that art of operating" and "art of thinking" (77). Stories provide, de Certeau says, the "decorative container of a *narrativity* for everyday practices," which "provide a panoply of schemas for action" (70). In other words, stories both describe and hypothesize everyday practices.

Second, stories connect us to each other as human beings. Telling stories is a process of knowledge construction that all humans share and in which all humans have some measure of competency because "we store, categorize, and process knowledge mainly in the form of narrative" (Bruner 1991, 4). In other words, we process and categorize knowledge in narrative form. Given this premise, consider the hammer. It is impossible to understand "hammer" without imagining it within a context of some kind. For me, the hammer is a symbol of my dad's identity—a master craftsman. I understand a hammer in this context: as part of the many tools that defined my dad's craft; as part of my dad's tool belt; as an extension of his hand as he built one of the many hutches or homes for which he was most known; as part of the many lessons about construction ("Hickory is the best wood for constructing hammers"); and as part of how we buried him—with his hands wrapped around the same hammer he used to begin his career. Although each of these parts could lead to a number of stories that explain better what a hammer is, the story most effectively describing its particular characteristics occurred during my dad's wake. After paying his respects, one of my dad's colleagues greeted us and said, "It's a shame to bury him with that hammer.

It was just getting broken in" (the hammer was then forty years old). This statement makes sense only when one considers its context: it made sense to us because to me a hammer is not simply a tool; it is a narrative construction of who my dad was.

In "The Narrative Construction of Reality," Bruner (1990, 1991) says that stories are "a form not only of representing but of constituting reality" (1991, 5) that work by constructing a dual landscape involving both consciousness (a way of thinking) and action (a way of operating), constructions that "occur concurrently" (1990, 51). Similarly, in *Acts of Meaning,* Bruner (1990) indicates that the human "capacity to render experience in terms of narrative is not just child's place, but an instrument for making meaning that dominates much of our life in culture" (97). He describes how the mental powers of narrative make it possible to frame experience in ways that enable us to both remember and make sense of human happenings"; in fact, he argues that "what does not get structured narratively suffers loss in memory." Narrative frames, Bruner says, provide a "means of constructing the world, of characterizing its flow, [and] of segmenting events within that world," without which we'd be "lost in a murk of chaotic experiences" (56). Human beings, he says, do not "deal with the world event by event or with text sentence by sentence," they frame events and sentences in larger structures" (64). However, simply reciting what happened does not constitute a narrative construction of reality because the "act of constructing a narrative . . . is considerably more than 'selecting' events either from real life, from memory, or from fantasy and then placing them in an appropriate order. The events themselves need to be constituted in light of the overall narrative" (1981, 8). This "part-whole textual interdependence" is a defining property of hermeneutics, because the "telling of a story and its comprehension *as* a story depend on the human capacity to process knowledge in this interpretive way" (8). This property is what makes narrative constructions of reality "different from logical procedures"—"they must be interpreted" (1991, 60).

Stories have more to do with context than with text, with the conditions of telling than with what is told. In other words, the events must be interpreted in order to tell the story. Hermeneutics is the study of interpretation, that is, how interpretation happens. Narratives have to do with people acting in situations. Making sense of a story (either in the telling of or listening to a story) requires making connections between characters' intentional states (beliefs, desires, theories, values, and such) and "the happenings that befall them." To make this connection is to state "reasons," not "causes," for behavior, a process Bruner calls *hermeneutic composability.* The term *hermeneutic,* Bruner says, implies an attempt to "express" or "extract" a meaning, which further implies that "there is a difference between what is *expressed* in the text and what the text might *mean.*" This hermeneutic process is required "when there is neither a *rational* method . . . nor an empirical method" for "determining the verifiability of the constituent elements that make up the text" (Bruner 1981, 7). Because interpretation is dependent upon an individual's ability "to achieve mastery of social reality," the "best hope of hermeneutic analysis," Bruner says, "is to provide an intuitively convincing account of the meaning of the text as a whole in light of the constituent parts that make it up," a process "nowhere better illustrated than in narrative" (8).

Narratives are not self-evident. They "do not provide causal explanations" for a character's actions; what they do supply is a "basis for interpretation," that is, a basis for "assigning meaning" to a text (Bruner 1981, 7). Events are meaningless without interpretation because the veracity of a narrative depends on the ability of the storyteller to situate a story within a context and to interpret the meaning of those events *based on* a particular context and to convince a listener to accept that version of reality. Because of this interlocutionary interaction, Bruner says, interpretation is "studded with" two problems that have to do more "with context than text, with the conditions on telling rather than with what is told": intention (purpose) and background knowledge (ability to judge veracity). Intention refers to the reasons a story is told, how and when it is told, and how it is interpreted "by interlocutors caught in different intentional stances themselves." Narratives (or their interpretations) are not created unintentionally: text, context, and situation converge to influence meaning for both the storyteller and the listener. Equally important is the background knowledge on which both the storyteller and the listener rely to judge the verisimilitude of a narrative account: typically, we presuppose that what an interlocutor says in replying to us is topic relevant and that we most often assign an interpretation to it accordingly in order to make it so" (10). Both these contextual issues hold "important grounds for negotiating how a story shall be taken . . . or how it should be told" (11), and both depend on the abilities of the storyteller and the listener to "fill in" information as necessary for comprehension (10). The capacity to complete information is the "human push to organize experience narratively" (Bruner 1990, 79).

A story is successful if it can convince listeners to accept its version of reality as "narrative truth"—if it can "sensitize us to experience our own lives in ways to match" (Bruner 1981, 13)—a truth "judged by its verisimilitude rather than its verifiability" (13). In this way, narratives are "centrally concerned with cultural legitimacy" (15); that is, they grow out of and reenforce cultural norms and encapsulate background scripts, implicitly inscribing the norms and behaviors of a culture. Because narrative is "centrally concerned with cultural legitimacy," stories not situated within a culture's norms seem "'pointless' rather than storylike" (11). But these scripts provide only the background necessary for comprehending the facts of the story; they do not constitute the "story" or its tellability. The "tellability" of a story depends on "what happened and why [a story] is worth telling" (12). To be worth telling, Bruner argues, a "tale must be about how an implicit canonical script has been breached, violated, or deviated from in a manner to do" harm to an implicit canonical script (11). In much the same way that de Certeau (1984) says an audience understands the *metis* component of storytelling, that is, the point of manipulation within the story that marks its unusualness, Bruner (1981) says that the moment a "hearer is made suspicious of the 'facts' of a story or the ulterior motives of a narrator"—an element of breach—she becomes "hermeneutically alert" (10). This state of mind comes from narrative necessity, which sets up the story in such a way that it "predisposes its hearers to one and only one interpretation" (9). These are the stories worth telling and worth listening to because they compel us into what Bruner calls "unrehearsed interpretative activity," or using what is known to understand what is unknown.

But, what does all this have to do with technical communication? How does a narrative way of knowing work in technical communication classrooms? For one thing, technical documents appear to be neutral, decontextualized texts that should require no interpretative activity (some scholars argue that this "objectivity" is the objective of technical documents—to limit interpretation; see Moore 1996, for example). Many textbooks focus on this aspect of technical communication, which I think doesn't address the content (how writers understand what they are saying) at a level in which students feel connected with the text. Stories, I think, do just that—connect with students at a level that all humans share. I don't simply tell stories about myself in classes. I assign a specific piece of literature as a context for assignments, which provides a way for students to make connections between what they already know—in a form they already know it (narrative)—and what they are learning about technical documentation.

Throughout the rest of the chapter, I describe how to structure a pedagogy designed as a narrative way of knowing, the procedures I use for helping students read and understand literature from a technical communication perspective, some individual and collaborative assignments I've used, and the evaluation methods used to assess student performance. I conclude this chapter with a discussion about some lessons I've learned from using this pedagogy.

ESTABLISHING A NARRATIVE WAY OF KNOWING

Course Focus

As a pedagogical approach, a narrative way of knowing begins with a theme (or focus), around which all assignments and discussions revolve (such as agriculture communication or environmental communication). This theme provides the focus for discussing and creating technical documents and depends on a teacher's own interests and goals for the class. The introduction of this theme should start with a representative technical document for a particular kind of communicative activity. The document could be a policy statement (government, corporate, nonprofit, or community), an application (such as for loans, admission, or adoption), a letter (such as Bush's recent letter to China), a proposal (such as for a national park or a legislative bill), a report (such as Accreditation Board for Engineering and Technology [ABET] 2000), and so on; there are endless possibilities. I generally start with a technical document because from a surface inspection, it appears to be neutral, objective, and decontextualized. Since adopting this approach, I have used such documents as the Agriculture Adjustment Act of 1933 (AAA), a piece of legislation that guaranteed farmers restitution if they planted only a percentage of their acreage, or the Environmental Protection Act of 1970 (EPA), which established the Environmental Protection Agency.

Literature

After selecting a technical document, I choose a piece of literature—or what I'm calling a narrative way of knowing—that provides a common context for thinking about the technical document. Providing a context for that document involves judiciously choosing a piece of literature that contextualizes the communicative activity implicitly embedded within the seemingly neutral technical document. The literature chosen should provide various perspectives (how characters think and interpret), situations (of collaboration), and actions (decisions made about communicative problems). This choice of literature could be one or more short stories, a novel, a film, a poem, or song lyrics—but whatever its genre, it must provide a comprehensive, complete story about a particular situation.

With the Agriculture Adjustment Act, for example, I used various short stories depicting farm life, values, and beliefs and a short excerpt from Lois Philips Hudson's (1984) *The Bones of Plenty,* which provides a lively scene in which North Dakota farmers attend a town hall meeting to discuss the merits of the Act with a government representative. With the Environmental Protection Act, I used Scott Russell Sanders's *Terrarium* (1985), a futurist novel depicting an overpolluted earth that forces people to move into Enclosures (globe-like structures that offer protection from the elements of climate) in order to sustain human life. Together, the technical document and the literature provide the context in which assignments and discussions revolve.

I use literature because its self-enclosed construction provides what Barbara Mirel (1998) calls "entry and exit points" that help students situate themselves within a context as a basis for interpretation, as a basis for "figur[ing] it out" (or as a student once described it—"it makes us use our minds"). The literature frames discussions about technical documentation in ways that situate students into what Bruner (1981) calls "unrehearsed interpretative activities" (9). Teaching technical communication as a series of parts without constituting them within the whole shortchanges students and encourages them to leave your classroom knowing only a particular skill—such as how to format a memo (placement of heading, formatting of body text, and so forth), which teaches them *what* to think, not *how* to think about communicative problems.[1]

Using literature as a context encourages students to consider thoughtfully the perspectives of characters in terms of the communicative action in the story as well as their and others' perspectives in the class.

In class, these perspectives are shared through in-class activities requiring students to discuss their interpretations in order to complete the assignment and through display of student writing on an overhead. For example, during an in-class writing activity, I asked students to work in groups of three or four to write parenthetical, formal, and expanded definitions for such terms as farm, agriculture, tractor, combine, homestead, and barn from the perspective of a character in Wil Weaver's (1989) "A Gravestone Made of Wheat" (such as the farmer, the farmer's wife, the sheriff, the farmer's son or daughter, the county clerk, the FHA agent, or the judge). My intention with this activity was to encourage students to pay attention to the implicit canonical scripts—embedded within the story and within students' personal narrative constructs—suggested by terms like "farm" and "farmer." By asking students to write technical definitions from a particular character's point of view, I had hoped they would expand their awareness of different contexts that affect the construction of knowledge and audience. In his end-of-the-quarter reflection memo, one student demonstrated this expanded awareness.

> I think that reading and discussing the required literature had an effect on the technical writing I did in this course. . . . I think that the biggest thing the reading did for me was to enlighten me on writing for different audiences. . . . As a specific example of this, I remember on one of our workshop days when we had to write technical definitions of certain farm-related terms for a character in "A Gravestone Made of Wheat." My group chose to write the definitions with the judge as the audience. I did not realize how much different it would be to write the definitions for the different characters. From doing the reading, I learned that the judge was well educated, knew little about farming, and seemed to have some biases towards farmers. These facts dramatically changed the way we defined the terms. We decided that we could use fairly technical terms to make the definitions because he must have been fairly well educated to be a judge, but we need to go into great detail in defining the terms because he did not have any experience in farming. The hardest part about defining the terms for the judge was trying to deal with the biases he had about farmers and farming. We ended up portraying farming in a somewhat negative way to make him better understand what we were saying because of his biases.
>
> Anthony (pseudonym)

To realize that language use changes with the audience and the situation, Anthony had to interpret the circumstances of the story to come to an understanding of the judge's attitude toward farmers; he had to, essentially, figure out what is implicitly provided about the judge. Anthony's expressive "I did not realize how much different it would be" statement indicates his engagement in "unrehearsed interpretative activity" in ways that challenged his conceptions about the objectiveness of technical documents. He clearly sees the judge, the term, and technical communication differently.

Although pedagogical design of a narrative way of knowing should begin with the selection of a technical document, decisions about the document and the literature more often occur concurrently. For example, *The Bones of Plenty* excerpt focuses specifically on the concept of agriculture adjustment, which led me to the AAA of 1933, and *Terrarium* mentions a fictitious "Enclosure Act," which led me to the EPA of 1970—both of which led me to the technical documents. Other possibilities include adoption policies of Native American children with Barbara Kingsolver's *The Bean Trees,* immigration policies with Helena Vermontes's *Under the Feet of Jesus,* a Search for Extraterrestrial Intelligence (SETI) report with Maria Doria Russell's *The Sparrow,* the Communications Act with *The Net,* or a NASA report with parts or all of the HBO miniseries *From the Earth to the Moon.*

For the purposes of this chapter, I focus my discussion on my use of the Environmental Protection Act of 1970 and Scott Russell Sanders's *Terrarium.*

Procedures

For literature to work as a context successfully, students need to understand its purpose in conjunction with technical communication. To help students situate the literature, in this case *Terrarium,* within the language of technical communication, I associate the communicative practices with those in the workplace by creating procedures for reading the literature (see appendix A). I use the term *procedures* for three reasons: *(a)* because it provides a lens through which students can view the context from the perspective of technical communication practices, *(b)* because it emulates the language and sensibilities of engineers and technical communicators, and *(c)* because the students at Michigan Technological University, where I was teaching at the time, tend to be extremely systematic and respond well to assignments that provide an identifiable foundation. The creation of these procedures grew out of my concern that because the assignment included a piece of literature, students might be tempted to adopt a literary studies perspective and read it in terms of its value as a literary artifact. I developed these procedures to help students focus their attention on the context the literature provides and how narrative ways of knowing transmit implicit knowledge.

Explanation of these procedures occurs on the day the schedule requires students to finish reading *Terrarium.* Column 1 (appendix A) names the procedure according to the cognitive function involved, column 2 provides heuristic questions that help students figure out the procedure described, and column 3 equates the activity with workplace activities. Although this discussion does focus primarily on *Terrarium,* when first explaining the procedures, I usually add to the discussion with examples from my work experience (such as when the small midwestern college I worked for in the early 1990s considered dissolving my public relations position, arguing that one marketing representation was enough for all six colleges in the region; in defense of my position, I was asked to write a brief statement about the value of this public relations position). After explaining the procedures, I ask student to work in groups of three or four, writing a one-to-two sentence statement for each procedure, using *Terrarium* as a context. I then ask each group to write one of their answers on the board and discuss it with the entire class.

The first procedure, Comprehending the Story, asks students to consider the meaning of the story as a whole. The biggest hurdle to overcome here is students' tendency to focus on the plot, such as "*Terrarium* is a story about how a group of people escape from the Enclosure" or "*Terrarium* is about good versus evil." The second procedure, Determining the Rhetorical Situation, asks students to consider the circumstances and the context of the story and how they affect the meaning of the story as a whole. Students also tend to describe the plot here, but this activity should focus on why the situation is important. Identifying the Exigency of the Situation, the third procedure, is the most confusing for students, mainly because of the term *exigency.* This procedure helps students identify the action or burden called for by the situation.[2] The fourth procedure, Identifying the Stakeholders, asks students to identify the people involved in the situation and the significance of their participation. This procedure emphasizes the importance of audience analysis in the creation of technical documents. The last procedure, Reflecting and Connecting the Story to Technical Communication, is intended to help students equate their interpretative activities with the activities associated with creating technical documents such as audience analysis, gathering and organizing information, and determining ethical dilemmas. This last step has proven to be the most difficult for students, I suspect, because this discussion occurs early in the term and because they try to see *Terrarium* as a technical document.

Assignments

To be successful, assignments based on the literature must address some kind of exigency—some reason or purpose—for providing the information that suits the context of the story. In other words, students should respond to a particular communicative problem not just answer a question with a particular form. Assignments require writing scenarios in which students more or less interact with characters. This interaction gives them a sense of audience in ways that simply naming "your boss" or "a client" cannot. It also allows them to imagine themselves acting in the situation. Assignments should not identify for the students the appropriate communication required (report, memo, or letter of application, for example), although there are some exceptions; rather, assignments should require students to interpret the situation and determine the appropriate action—to make rhetorical decisions. Appendix B lists some representative assignments I have given in conjunction with *The Bones of Plenty* and *Terrarium.*

What's important to keep in mind when creating assignments are the connections the students make between this imaginary "playacting" and the kind of communicative interactions in which students will be expected to participate in the world of work. These assignments must involve appropriate and recognizable workplace situations, actions, and contexts with which students connect the rhetorical action—or exigency—between what they know or have been told about the workplace, whether that knowledge is from personal experience, from professors in major classes, or from professionals in the fields, and the kind of thinking and writing involved in the assignments.

I generally focus the first part of the term on individual and in-class collaborative assignments, which are intended to help students practice writing in the genres of technical communication and understanding those "genres as social action" (Miller 1984) and as constructions of knowledge (Berkenkotter and Huckin 1995). During the last half of the term, students work in collaborative groups of four or five on a major project in which they emulate the contextual approach I demonstrated during the first half. The purpose of the major project is, obviously, to evaluate students' abilities to generate content for a specific audience, to organize and shape that content, and to present this information in a readable fashion—to assess their communicative competency.

Individual Assignments

Assignments like those listed in appendix B are individual assignments that take place during the first half of the term. They all rely on the story as a context and require that students interpret a character's intentions, motivations, and actions. Assignments involve two kinds of communicative activities—individual and collaborative—both of which provide students with opportunities to practice articulating what they know, how they know what they know, and why they know what they know.

Pragmatically, these assignments involve creating documents such as letters of application and memos, fact sheets, short reports, technical descriptions, and processes and procedures. On a more conceptual level, to create these documents, students must interpret the circumstances embedded in the story in order to generate the content for the documents, contextualizing and organizing that content within the constraints of the assignment and the audience. These interpretative activities include topics such as identifying and characterizing problems, situations, and actions; analyzing and assessing problem-solving strategies and reflecting on those problems; and describing and evaluating collaborative processes, procedures, and instructions. Assignments based on *Terrarium,* then, are constructed in such a way that students are required to make genre decisions, to generate content, and to format and organize information—to conduct themselves as a member of that community—based on their assessment of the situation, the audience, and the exigency of the assignments. Although the technical forms assigned are fairly standardized (for example, memos, letters, reports, and proposals), the content for those forms must come from students' interpretations of the circumstances of the story. In this way, students not only practice the interpretative task of assigning meaning to an event, but they also unmask the nature of practice within a community.

To complete any of these assignments, logistically, students must make genre decisions, invent content appropriate to the situation, and present this information in an appropriate form, style, and tone. These decisions require students to identify the rhetorical situation (such as a second interview or an expert review), the exigency (what's required to get a particular position or request for evaluation, for example), and select the appropriate form, content, and style, based on this situation (for instance, applying for a job requires a formal letter of application or pitching an idea requires a proposal). More conceptually, students must interpret the circumstances of the story, invent content from their interpretations, and organize that information in ways that make sense to the audience defined—all in terms of the purpose and exigency of the assignment. To do this kind of conceptual thinking, students must see themselves practicing imaginatively within the context of the story.

To be effective, individual assignments based on the literature need to be contextualized within some kind of exigency—some reason or purpose—for providing this kind of information. In other words, the assignments should provide a scenario in which students are responding to a communicative problem, not just answering a question. One assignment, for example, asked students to apply for a position with a character from *Terrarium:*

> You have applied for an engineering position with The Enclosure Group by answering a blind ad in the *Enclosure Gazette.* So far, you've had one introductory meeting with several Enclosure representatives and feel confident that they liked you. You received a letter today from Dr. Zuni Franklin, the Supervising Engineer, at 3980 Enclosure #1, Portland City, CA 00001, indicating that you are one of five applicants competing for the job. She has asked you to respond to the scenario below in writing to determine if you will be called for a second interview. Using *Terrarium* as a context, write a letter to Dr. Franklin indicating your continued interest in the position and identifying, analyzing, and evaluating three problem-solving strategies from the list below. Also indicate which strategy you think works best and why.
>
> Death of Sol
> Phoenix's fear of Terra
> Avoiding the health patrollers
> Repairing the enclosure
> Zuni's retirement

Teeg and her father
Teeg and her mother

In a related example, but not nearly as successful, I asked students to use *Terrarium* as a context and to create a fact sheet intended to convince people like Judith Passio (a known adversary of the Enclosure—the globe-like structure into which humanity was moved when the earth became so polluted that it could no longer sustain life) to accept the inevitability of enclosures. In the novel, Passio is a holdout, refusing to move into the enclosure throughout the story and, aside from twenty or so pages toward the end of the novel, Passio's character is known only from diary-like vignettes between chapters. Passio is, as a lawyer might surmise, a hostile witness (or audience), firmly believing in her rejection of technology as the sum total of humanity's problems—the enclosure representing the furthermost extent of this problem. To write this fact sheet, students needed to interpret her character as hostile in order to use language that could actually convince her to move into the enclosure. Although most students wrote "effective" facts sheets from a technical communication perspective in terms of clarity, organization, and design of information, most of them did not consider the reality of Passio's character in their use of language. Many of them used a "let's-be-friends" voice, highlighting the benefits of enclosure life, many of which Passio had openly criticized.

Although most of the assignments typically succeed (that is, students write clear, effective documents based on the context of the story), when they do fail, it is not always the students who demonstrate bad judgment. With the application letter assignment described earlier, students generally accurately addressed a letter of application to Dr. Gregory Passio and competently identified, described, and analyzed three problem-solving strategies. However, they also referred to events that this character couldn't possibly know because he was long since dead when they occurred. Because the students had seemingly understood the idea of literature in the technical communication classroom and had successfully integrated content from the story in previous assignments, I was confounded that in their character analysis, they had missed such an obvious point. I didn't expect to have to discuss character analysis in this way. So, I asked students why they referred to things the character couldn't possibly know. One student raised his hand and said, "Well, all the scenarios you listed for the assignment happened after his death. We just assumed you brought him back to life for the assignment." Evidently, I had.

Collaborative Assignments

The major project requires students to work collaboratively in groups of three or four, asking them to create a microcosm of their field of study through the lens of a technical document indigenous to the field. Some groups, however, involve more than one major. The technical document created should represent some larger concept indicative of that field (the ethics of artificial intelligence, for instance). Students must figure out a focus for their projects, whether the group consists of similar or dissimilar disciplines, in ways that bring the three fields together, such as a piece of technology (a transistor) or a concept (project management). And because a group's complement does not always consist of students with the same majors, they must create a document that represents two or more fields. They then must choose a story that provides a context for the technical document they are creating, in much the same way that I use *Terrarium* as a context for the Environmental Protection Act.[3] Within these constraints, each group completes a number of ancillary assignments intended to broaden their knowledge and understanding of the concept through research and development. These ancillary assignments include a proposal, a journal report, an audience-analysis report, a visual report, and individual and group activity logs.

The audience for the assignment consists of technical communication teachers who want to know more about the different disciplines on campus in order to better teach technical communication. This major project includes a variety of ancillary assignments that inform their thinking about the content of the document: a proposal, a journal report, an audience-analysis report, two progress reports, a cover memo, and a final presentation. These assignments accompany and support the technical document students create. Appendix C shows some representative projects students have completed during the past four years.

As part of this assignment, students are expected to include, either within the project itself or as part of the appendices, presentation materials and a summary and analysis of the story and how it works as a context for their document. Although I was concerned in the beginning that I'd have to provide too much input in the selections of stories, my input has been minimal. I was especially worried that students would have difficulty successfully articulating the connection between the document and the story. Although some groups tended to engage in plot summary more than interpretation of the story in the context of their project's focus, most groups competently analyzed the story from the perspective of their project's focus. One group that called itself the AI Group, for example, effectively connected the chosen story *(Bladerunner)* to their topic—artificial intelligence—by drawing on their background knowledge in computer science.[4]

> Artificial Intelligence, being a relatively new area of study (middle of 20th century), has largely extended it's interest into a variety of different disciplines. It can be loosely defined as: the quest to understand thought patterns and recognition processes present in the mind of living organisms, and to somehow reconstruct these thought processes in such a way that a machine (computer) can mimic parts or all of the process attained in living thought. Such a philosophical definition leaves much unanswered, as is the case in Artificial Intelligence.
>
> Considering what intelligence is, AI researches have found much resistance in modeling brain thought and learning. Often, even simple tasks which most all people can achieve without much "intelligence", prove to be large obstacles in AI. For example, most people have the ability to clean the dinner table and do dishes after a meal. Such a task is a large problem for an Artificial Intelligent machine to complete without assistance.
>
> Such implications lead us to believe a living brain uses more than intelligence for many everyday applications. Extending intelligence to include such things as feelings or environmental awareness in contextual situations (a common unconscious process in the human mind), drives AI research to include such things as body and natural language. These topics have become much more complex than originally projected, and thus have driven the interest of Artificial Intelligence into many areas of study.

They also include links to Web pages that provide more background information and list resources for future reference. In connection to this background knowledge, they clearly indicated how the focus of their project connected to their story by emphasizing the ethical considerations involved in development of this kind of technology:

> In a number of ways computer science and engineering are like the law profession, all three rely on precedents to make decisions. A computer scientist always wants to go into new situation prepared by a prior precedent.

> Unfortunately in the rapidly developing field of computer science, precedents are not available, or they are of limited use. The solution to the short fall is the world of fiction. Fiction allows the computer science community to explore the ethical implications of their work, even if the necessary advances will not be available for years to come. By creating an ethical precedent, the computer scientist will be prepared for the road ahead, and he/she will be unhindered by the limited existing "real world" precedents. The ability to be forward looking makes fiction an extremely important aspect to a well rounded Computer Scientist.

This group identified fiction—specifically science fiction—as one way to fill the gap created by unavailable precedents in the fairly new field of computer science. Because science fiction often depicts computerized societies, it creates, according to this group, an "ethical precedent." By focusing on the ethics of artificial intelligence, they attend to the contexts relevant to such a technology and address the humanistic components that interest technical communication teachers. Interestingly, they did not ignore their second audience—other students in the class who would be in attendance for their presentations. In both cases, these students addressed attitudes, motivations, skill, education, and interest of their audience in order to "relate the presented material to the audience on a personal level."

Similarly, a student from an early class focused his report on the idea of progress by tracing the history of the engine from combustible to fuel-injected.[5] For context, this student selected the jalopy in John Steinbeck's *The Grapes of Wrath* and its Western theme so he could show the "struggle between human and machine." He also placed his discussion within the larger context of the conflicts between humans: "While the struggle between human and machine was a relatively simple battle, the struggle of humans against nature and the battle between humans proved to be much more difficult to win." He defended his choice in literature appropriately by saying that the novel "portrays how humans struggle with machines" (the Joads are forced to repair their engine along the way to California), as well as how humans have struggled against nature (the Joads lost their land in the Dust Bowl). He argues that the Joad's won this struggle because "they were ultimately able to repair the engine and continue along the road," which represented their future. He associates this struggle with what he sees as the aim of mechanical engineering, that is, "to solve problems that deal with humans against machines," and demonstrates that connection by illustrating and describing the dynamics of a fuel-injected system. By contextualizing the theme of progress within *The Grapes of Wrath* and the Western narrative of progress, he demonstrates for teachers one of the underlying struggles important to mechanical engineers.

Another group focused on the role of regulations in the field of civil engineering. To do this, they characterized the building of the Hoover Dam, "when regulations concerning the environment [were] nonexistent," as an impossibility today in light of "public awareness." As a context, they pointed to Upton Sinclair's *The Jungle,* which led to the "passage of the Pure Food and Drug Act of 1906, less than a year after the novel's publication." They equated the way *The Jungle* "illustrated how changes in society's perspective can lead to changes in regulations" with the necessity for engineers "to understand how to identify and analyze. . . public opinion, environmental laws and regulations." The connection between their historical example of Hoover Dam and *The Jungle* shows how today's civil engineers must pay more attention to factors such as public opinion, preservation, and environmental protection "prior to construction"—all which help technical communication teachers better understand the nature of work in civil engineering.

The major project works much like the individual assignments in that both operate from a narrative way of knowing. Individual assignments help students learn to identify, describe, and evaluate practices from a critical perspective. They require interpretative acts in the construction of knowledge, encourage connections through contexts, and enable articulations of knowledge in a recognizable form. With the major project, students have an opportunity to engage in, as de Certeau (1984) says, a "narrativizing of practices" that encourages them to consciously consider what work means in their field, how it operates in that field, and their role in completing that work. The triangulating aspect of the major project—the bringing together of the field (or fields), the issues (journal), and the story, or as de Certeau might say, their *art* (its theory and practice)—demonstrates their ability to engage in narrative ways of knowing. The successful projects demonstrated what Bruner (1981) called a story's "verisimilitude," that is, the story's tellability, what makes it worth telling" (13). Their ability to "assign meaning" to their work through narrative constructions included well-designed, content-rich documents that illustrated their competency—their alertness—in conveying technical information to an audience, while engaged in unrehearsed hermeneutic activities.

CONCLUSION

Instructors adopting a narrative way of knowing as a pedagogical approach need to know that using literature in the technical communication classroom is not necessarily new (see, for example, Kilgore 1981; Karis 1989). These scholars argue, and I agree, that literature can provide both examples of and a context for technical communication; however, they do not encourage the production of technical documents out of the context literature provides. Teaching technical communication as a narrative way of knowing does just that: it not only provides opportunities for helping students develop an understanding of technical information as constructed from a context but also encourages reflective and critical perspectives about that information.

When choosing literature, instructors need to consider carefully whether the literature chosen is conducive to the construction of technical documents. The story should contain examples of collaborative activities, demonstrating limited and full participation and various levels of conflict and cooperation. The action depicted in the story should involve several aspects of people working together, of negotiation of meaning, and of application of that negotiation to a problem. The mutual participation depicted in the story should involve characters trying to figure out the circumstances of their lives, their work, and their world in conjunction with other characters within a context of practice. Stories that are more character-driven might not be able to demonstrate as effectively the kind of mutual participation necessary to engage students in the practices of that community.

If adopting this approach, instructors need not be rhetorical, literary, or narrative theory experts, although they should be able to explain how a particular story provides a context for a particular technical document. When instructing students about the use of literature, instructors should adopt a rhetorical criticism approach, rather than a literary analysis approach, because its emphasis on audience, purpose, and situation better fits the socially constructed theories common today in technical communication discussions. Instructors might find it useful to read current theories about how communities of practice use narrative ways of knowing to sustain relationships within a community (see, for example, Wenger 1998).

More than anything else, I feel compelled to warn instructors adopting this approach that preparing for class and assessing and evaluating course documents can be time consuming, at least in the beginning. Creating scenarios for assignments requires a great deal of creativity: you must situate students within the context of the story in ways that require them to act (through the creation of a technical document). You must also be prepared for the various interpretations students present. I've used this approach for almost five years and inevitably a few students will write on their evaluations that they had difficulty figuring out what I want. These statements come from, I think, their understanding of my written comments on their documents as arguments with their interpretation (rather than as a response to the clarity of their statements).

One technique I recently adopted has helped me better help students evaluate and assess their work in my class. Before completing assignments, I have students engage in an assessment workshop—a method for evaluating technical documents holistically. This method encourages them to exercise their judgment—*before completing an assignment*—by evaluating a technical document according to an assessment rubric that I include in the course materials. In this assessment workshop, the instructor's role is one of guide; she does not score or critique the document herself. Although I have only just begun using this method, I am already discovering that students generally identify the same problems with a document that I would, that students are much more critical than I would be, and that students write better and require fewer revisions because they have an idea about how their documents will be evaluated. Overall, I have found that the longer prep times for course discussions, as well as for the assessment workshop, pay off in time spent responding to and grading students' documents.

Certainly, I can't prove that telling and using stories make students better writers or communicators. I simply believe that stories are a more interesting way to learn. When I run into students after the completion of a class, they inevitability remember *Terrarium* before they remember my name. They always remember the stories.

APPENDIX A

PROCEDURES FOR THINKING ABOUT LITERATURE IN THE TECHNICAL COMMUNICATION CLASSROOM

Competency	*Heuristic*	*Workplace*	*Knowledge and Skills*
Comprehension: What's it all about?	What is the story about? Imagine that you must tell someone else what it is about and you only have one minute. In one sentence, summarize the story. This summary should grasp the essence of the story, that is, how you interpret the overall significance of the story from a particular perspective.	Identification of tasks, activities, and problems.	Summarizing Paraphrasing
Description: What is the Situation?	What is the context/ What are the circumstances of the story? In other words, what's going on? Describing the "context" should go beyond the mere plot (what happened—the events) of the story. It should indicate the relevance of the context to the communicative action.	Identification and articulation of the contexts and constraints affecting a situation.	Analyzing Problem-solving
Configuration: What needs to be done?	What is the exigency of the situation described in the story? What is the communicative action, burden, or problem?	Determination of what kind and form of communication is required in a particular situation.	Evaluation Assessment Decision-Making
Categorization: Who's involved?	Who are the main players and what part do they play in the action of the story? What is their significance to the situation and the exigency?	Identification of the people involved in the communicative action.	Categorization Identification
Reflection: What do you think?	What do you think now that you have thought about the story in these terms?	Figuring it out; telling why, stating reasons.	Critical Thinking Synthesis

APPENDIX B

SAMPLE INDIVIDUAL ASSIGNMENTS

Goals & Objectives	*Assignments*	
	Agriculture Adjustment Act of 1993 *The Bones of Plenty*	Environmental Protection Act *Terrarium*
Address the rhetorical situation and appropriate audience	In a memo to Mr. Petersen, a North Dakota farmer considering supporting the AAA, summarize the AAA and *The Bones of Plenty* and indicate any connections between the two "stories."	You have recently been hired by The Enclosure Group as a supervising engineer. Dr. Gregory Passio has requested that you evaluate three problem-solving strategies used by various teams of engineers in *Terrarium* as your first task. Identify, describe, and evaluate, compare and contrast them, and indicate which one you think works best and why.
	In the Netforum discussion for today, discuss why you think the FHA and the farmers had such difficulty communicating with each other. Related in-class activity: In groups of three or four summarize these statements and be prepared to discuss the assumptions underlying the statements (e.g., stereotypes about farmers or government agents).	As a consultant hired by The Enclosure Group, you have been asked by Dr. Zuni Franklin to assess their collaborative processes and make recommendations to the board for improvement. She does not have time to meet with you personally and would like to have your evaluation in writing. In this evaluation identify, describe, and evaluate two collaborative processes and indicate which one you think is best and why in an appropriate form.
	The Judge reviewing Olaf's proposal for marriage to the German girl has asked for some clarifications of his language. In groups of three or four, respond to the Judge's request on Olaf's behalf, defining the following terms (farm, agriculture, farmers, etc.).	Imagine that we have a hearing-impaired student who is accompanied by an interpreter. The purpose of the interpreter is to translate what is said/going on in the class to the student. It is not the interpreter's job to read class material. Because we can't expect this interpreter to read *Terrarium* in order to accurately translate class discussion, we need to create definitions and descriptions of terms specific to Terrarium. Form groups of 3/4, choose one of the terms below, and write a formal, informal, and operational definition for that term. (Terms: ingathering, enclosure, pedbelt, wildgoers, terrarium, chemmies).

Manage the appropriate genre for the rhetorical situation	An FHA agent in charge of getting the news out to farmers about the AAA in an effort to convince them to support the bill. Create a fact sheet for farmers outlining the provisions of the AAA to be used at the Town Hall meeting next Friday. In preparation for her move to the US, Olaf's wife has requested more information about farming, farming life, and agriculture in general. Choose one of the items below and write and design a technical description to send to her: organizational structure of the US Department of Agriculture, the purpose of the Agriculture Extension Service, directions to Mr. Petersen's farm three miles north of Harvey, ND from Germany, the operation of a combine, the logistics of a windmill, or the purpose of the Morrill Land Grant Act.	As a representative of the Enclosure Group, Dr. Passio has asked you to create a fact sheet informing the residents of Earth of the procedures for moving into the Enclosure. In addition to the procedures, this fact sheet should also include contact information. Your co-worker, Phoenix Marshall is trying to explain to Judith Passio, who has never been in the Enclosure, how certain mechanical devices work. He has asked for your help. In the appropriate genre and tone, choose one of these processes (vaporizers, dismantling cities, pedbelts, eros and chemmie parlors, or establishing Jonah Colony) describe how its works. Your description should include a graphic of some kind.
Write in the appropriate register for the audience and with the right tone for the situation	Technology biography—Choose a character and describe her/his relationship to technology, how that relationship reflects her/his world view, and reflect on that choice as indicative of your own relationship and view of technology.	You have applied for an engineering position with The Enclosure Group by answering a blind ad in the *Enclosure Gazette*. So far, you have had one introductory meeting with government representatives and feel confident that they liked you. You received a letter today from Dr. Zuni Franklin, the Supervising Engineer, at 3980 Enclosure #1, Portland City, CA 00001, indicating that you are one of five other applicants competing for the job. He has asked you to respond to the scenario below in writing to determine if you will be called for a second interview. Using *Terrarium* as a context, respond to Dr. Franklin, indicating your continued interest in the position and identify, analyze and evaluate three problem-solving strategies from the list below. Also, indicate which strategy works best and why. Problems: Death of Sol, Phoenix's fear of Terra, Health Patrollers, Repairing the Enclosure, Zuni's Retirement, Teeg and Phoenix's relationship, Zuni Franklin and Judith Passio, Teeg and her father, Teeg and her mother.

APPENDIX C

SAMPLE COLLABORATIVE ASSIGNMENTS

Goals & Objectives	*Assignments*	
Create a logical and explicit arrangement that fits the readers, the genre, and the situation	Mechanical Design Area: Mechanical Engineering Focus: Safety Story: "Out of This Furnace" (Dobrejcak) "I choose this story because it illustrated my points in a very personal way. When talking about topics such as mechanical design, it is very easy to get lost in all the technical language without thinking of how the people who use the design are affected. The characters in the story died as a result of poor design of furnaces in the steel mill." (Anthony)	Forestry Area: Forest Management Focus: Sustainability Story: *The Wolves of Isle Royale: A Broken Balance* (Peterson) "The relationship of dependency between the wolves and moose is a good analogy to humankind's relationship to forestry. The same concept exists as not only do humans depend on good forests, but also the existence of good forests depends on human actions." (Matt, John, Tom, & Jason)
Select and invent content appropriate for the reader and the situation	Engine Design Area: Mechanical Engineering Focus: Progress Story: *The Grapes of Wrath* "*The Grapes of Wrath* is actually a wonderful selection because it portray how humans struggle with machines (when the Joad's are forced to repair their engine along the way to California), as well as how humans have struggled against nature (the Joad's lost their land in the Dust Bowl." (Matt)	Usability Testing Area: Scientific and Technical Communication Focus: Transparency Story: *Brave New World* (Huxley) "The society in *Brave New World* as a whole runs on transparency that goes unchecked. . . . They attempt to gain control by 'pressing the buttons' of those unconscious learned opinions burned into the mind of civilization." (Sara & Curtis)
Employ sound principles of layout and design in the creation of the finished document pages	Firewalls Area: Information Systems Focus: Security Story: *ruthless.com* (Clancy) "What this story did for me was to give me a way to relate my classroom work with actual outside occurrences that could or may be happening in the work world" (Lucas).	Hoover Dam Area: Environmental Engineering Focus: Regulations Story: *The Jungle* (Sinclair) "*The Jungle* provided valuable information for our project because it illustrated how changes in society's perspective can lead to changes in regulations. A novel, report, or article can greatly impact peoples perspective on an issue, leading them to pursue change and force legislation." (Jason, Andrew, Lindsay)

8

HYPERMEDIATING THE RESUME

James Kalmbach

INTRODUCTION

Popken (1999) suggests that the resume as a genre was codified in the 1920s and 1930s in various business communication textbooks, becoming part of technical writing textbooks in the late 1970s (95).[1] Today, resume writing is an assignment deeply woven into the technical writing curriculum. Most contemporary technical writing textbooks include a unit on letter and resume writing in some form, as do most classes.

The appeal of resumes (and letters of application) as a beginning assignment in a technical writing class is not hard to see. Many students approach such a course with a good deal of apprehension. Often they have done little substantive writing since first-year composition and have at best a limited sense of the role written communication will play in their professional lives. These students realize that resumes are important documents and are usually motivated to work on the assignment. The appeal to teachers is also not hard to see. A resume is an enormously flexible form. It is a brief, compact document that can be used to foreground virtually any theoretical or practical aspect of technical writing.

INNOVATIVE PRACTICES AND THE RESUME

Although the resume continues to be an attractive assignment, much has changed since the late seventies. Indeed, the technological and social contexts of resume writing have changed dramatically in just the last few years. The resume assignment in a technical writing course is in need of innovation to align it with current practice.

One of the goals of a traditional resume assignment has been to produce a resume that could be skimmed on a first reading by someone looking to separate candidates who clearly are not qualified from those who deserve a more careful look (McDowell 1987).[2] Traditionally, a job candidate could not count on more than ten to twenty seconds of a human resource person's time during this initial screening.[3] Faced with a stack of two hundred resumes and a limited amount of time to identify a pool of promising candidates, what choice was there? Screening to eliminate candidates first was and still is a form of survival.

In the late 1980s, one of the attractions of desktop publishing technology was its ability to transform a typewritten resume into an easy-to-scan typeset document. Today, a visually attractive typeset paper resume is still an important component of a job search, but at the same time, students are increasingly asked to email resumes or to submit resumes using Web forms that strip out all formatting. In addition, in many large companies, computers now do the task of initially screening candidates. A large insurance company headquartered near my university routinely scans every print resume it receives using the software package Resumex. They place the scanned resumes into a database so that initial pools of promising candidates can be generated via keyword searches.

In preparing a print resume for Resumex scanning, less is no longer more. Because an overworked human resources (HR) person is no longer doing the initial screening, job candidates can cram their resumes with text containing the keywords that they hope are likely to generate hits and help them make the initial cut. However, these job seekers still need an attractive, traditional paper resume to share with the (for now) human manager who actually interviews them.

Finally, the World Wide Web makes it possible to produce a resume that can be both scanned quickly and linked to increasingly more complex detail about a job candidate. Such a resume/electronic portfolio can function at many different levels and meet a variety of purposes.[4]

HYPERMEDIATING THE RESUME

These changes suggest that a single print resume is no longer enough. In this postmodern world of fragmented identity, students need a hypermediated resume that exists in several different versions designed for different media and different purposes. The resume, of course, has always been a hypertextual genre. Its stylized, scannable, nonlinear nature invites multiplicity in the sense that Bolter and Grusin (1999) have defined it. The symbiotic relationship of a resume to a letter of application also reflects Bolter and Grusin's notion of "immediacy"—the idea that through media we strive to create the illusion that the reader/viewer is part of the experience rather than separate from that experience. In a job search, the need for immediacy is filled by the letter of application in which the job applicant interprets the resume and attempts to make him or herself come alive for the reader, while the resume complements this immediacy with a nonlinear, selective, hypermediated explication of that background. The letter of application and the resume are deeply intertwined, just as immediacy and hypermediacy are intertwined, each depending on the other.

In the hypermediated resume assignment, not only do the print letter and resume play off one another, but different electronic versions of the resume further complicate the assignment. Specifically, this assignment repurposes the traditional paper resume into three (and sometimes four and five) different documents:

A more or less traditional print resume that uses visual form, space, and typographic emphasis to create a resume that a human can scan in ten to twenty seconds to identify a candidate's strengths

A text-only resume (with line lengths limited to sixty-five characters) that has no formatting and is suitable for submission via an email message or a Web-based form

An HTML version of the resume published on the Web

In addition to these three required versions of the resume, students may optionally submit a scannable version of their resume, create an Acrobat PDF version of their print resume, and do a more complex, HTML-based portfolio resume as a project later in the course.[5]

The Print Resume

The initial document in this project is a print resume pretty much like the print resumes students have always produced. No innovation here. Large companies may be moving to Web-based forms and resume databases, but most small companies still sort through resumes by hand. A clearly written, attractive, effectively organized resume is as important now as it has always been. Scanning software may be able to process densely-packed resumes full of keywords, but managers will always prefer a lean, focused, attractive resume that tells a story.

If there has been any change in the design of print resumes, that change has been toward more visually conservative forms. Because design elements such as graphics, rules, tints, and unusual fonts limit the effectiveness of OCR software, many students elect to avoid them.[6] A

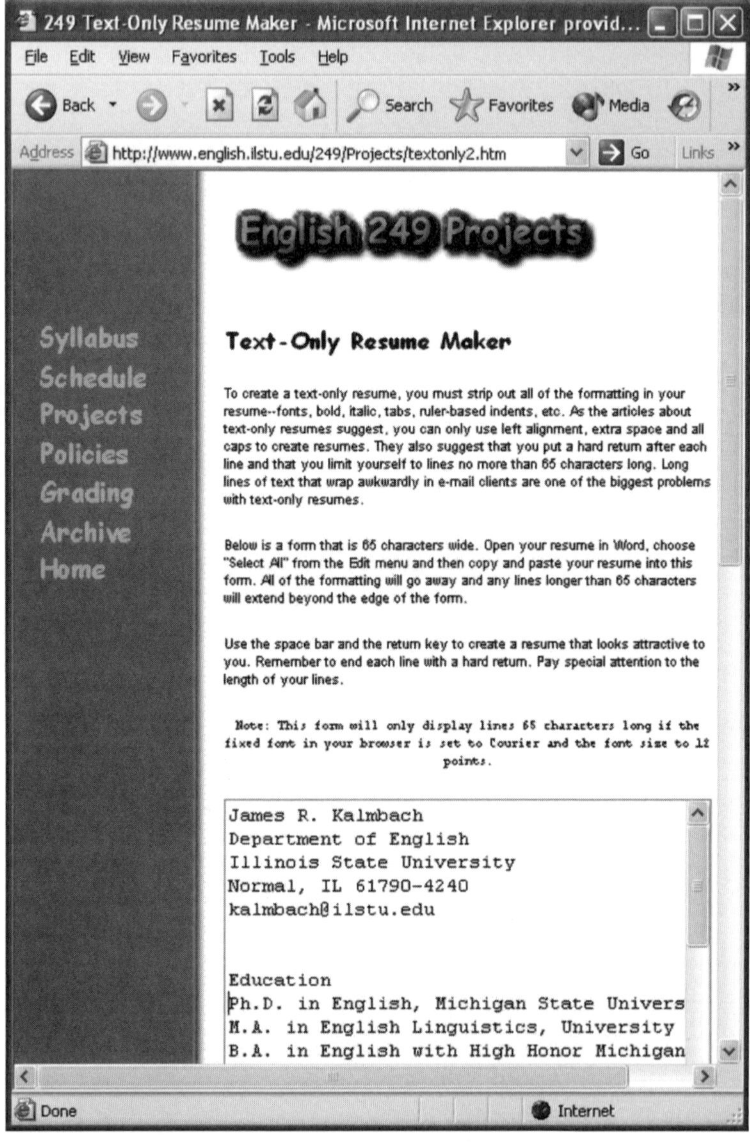

Figure 1. Web form used to create a text-only résumé. Lines that extend beyond the edge of the form will not wrap correctly when emailed. Available: www.cas.ilstu.edu/english/249/Projects/textonly.html

resume with a simple, straightforward visual design can, however, still be a complex rhetorical document; giving up fancy tints and rules does not mean giving up design.

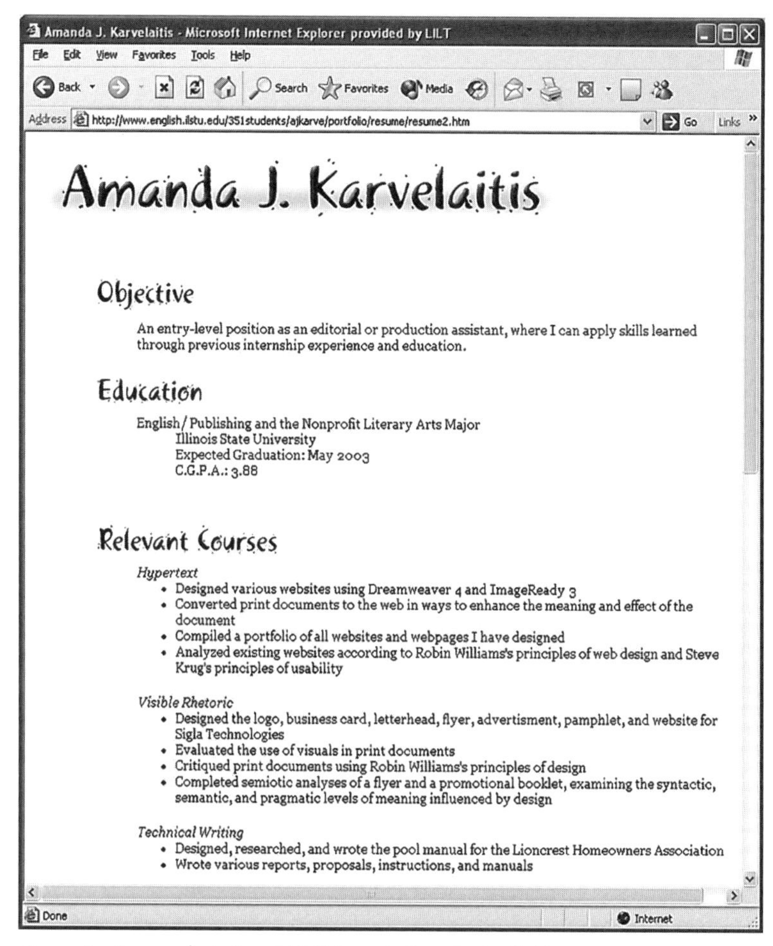

Figure 2. Screen-shot of an HTML-based résumé. Available: www.english.ilstu.edu/351students/ajkarve/portfolio/resume/resume2.htm

The Text-Only Resume

As more and more job recruiting takes place over the Internet, students are increasingly asked to submit resumes via Web-based forms that support only ASCII text, or they may be asked to submit a resume via email. Currently, the only reliable way to submit a resume by email is to paste a text-only version of the resume into the body of a message. Sending resumes via attachments to email messages can be a nightmare of incompatible word processing formats; the person processing the resume may not have the right fonts in his or her system to render the resume correctly; and the resume file may inadvertently carry macro viruses.

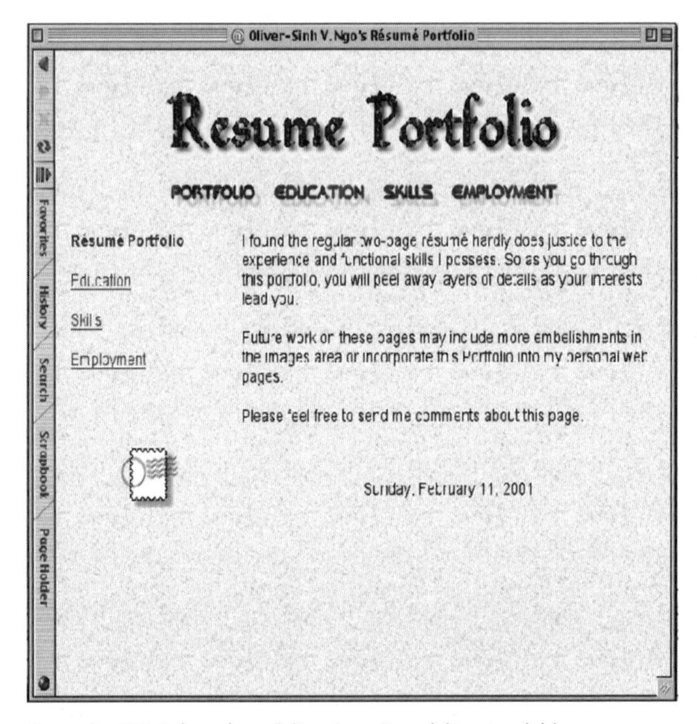

Figure 3. HTML-based portfolio résumé/portfolio. Available: www.ilstu.edu/~ovngo/portfolio.htm

Indeed many companies routinely delete without opening any resumes sent to them as attachments (Nemnich and Jandt 1999).[7]

Teaching students to create a text-only resume is harder than one might think. Even when students save their resume as text-only files in Word or use a program like Notepad, they still have access to fonts and characters not supported in Web forms. In addition, most word processing applications use word wrap, resulting in lines of text much longer than the sixty-five- character lines displayed by email programs. As a result, a text-only resume might look fine when viewed on the screen or printed from Notepad but turn into an unreadable jumble when pasted into an email message.

Creating a text-only resume in a mail program can also be dicey. Although Web mail clients tend to be safer choices, mail clients such as Eudora, Outlook, and Outlook Express support HTML formatting within mail messages. As a result, students can add formatting to their email without being aware that they have done so, and the resulting resume will be filled with HTML code that at best makes the resume hard to process and at worst makes it incomprehensible if a reader's email program does not support HTML encoding and instead shows the message with its underlying code.

My solution has been to use a Web-based form for converting Word documents into text- only resumes. The form in figure 1 is set to a width that displays sixty-five characters of a normal-sized nonproportional font.

To use the form, students open their resume in Word, copy the text, and paste that text into the text area. All non-ASCII formatting disappears, and any line that extends beyond the edge of the form must be shortened because it is too long to display properly in an email program. Students use returns to shorten those lines, and they add space between sections and upper-case headlines to create more emphasis. They then email me the text-only resume, and we go through several rounds of revision.

HTML Resumes

Because I also teach my department's Web-authoring class, I have always been on the lookout for ways to incorporate more Web-based publishing into my other classes. I began including an HTML resume in technical writing courses as a way of doing more with the Web and to give students the confidence to attempt a Web-based project later in the class, but I discovered that an HTML resume can also be part of a job search. Many students report receiving inquiries as a result of having their resume online. Moreover, the Web is a medium in which students can experiment with design elements such as color, line art, and even photos, not typically available in print resumes (see figure 2).

Publishing on the Web can also expand students' notions of what is possible in a resume. An HTML resume can both tell a story in ten to twenty seconds and provide links to more detailed information in different topics. For example, in the HTML resume pictures in figure 3, the links along the left side of this resume lead to levels of increasing detail about the student's education, skills, and work experience. This sort of resume is both scannable by a human and packed with information should the reader wish to know more.

Whereas the text-only resume is an exercise in minimalism (students explore just how much formatting they can strip away from their resume) the HTML resume is an exercise in expansion. Students experiment with the different ways they can use digital media to add design elements to their HTML resumes while still telling a compelling story.

INTEGRATING THE HYPERMEDIATED RESUME INTO THE TECHNICAL WRITING CLASSROOM

Teaching so many different forms of the resume has also meant innovating the ways the assignment is sequenced. Instead of teaching a single unit in a three-week block of time, I teach the unit in two parts: at the beginning of the semester and again after midterms. The initial week is spent introducing the assignment, workshopping first drafts of letters and resumes, and talking about formatting issues. After this week, we turn our attention to writing first a report and then a manual. While working through these projects, students turn in a new draft of their letter and resume each week until those resumes are clearly written, attractively formatted, and error free.

After this period away from the project, we return to the hypermediated resume. We spend one day converting print resumes to text-only resumes and the next five class periods converting text-only resumes into HTML resumes.

Using a discontinuous structure has been particularly helpful in this assignment. It gives students the time to work through various writing and design issues as they create an effective print resume. Once that resume is done, they can turn to repurposing it in other media.

Although the assignment as described here is closely tied to my particular institutional context and my style of teaching with technology, the hypermediated resume can be used in a variety of pedagogical contexts. Successfully incorporating such an assignment will depend on three issues: the nature of your students, the technological resources available, and the pedagogical goals that you bring to your courses.

The Nature of Your Students

The single most important factor in deciding whether to teach the hypermediated resume in a technical writing course is the nature of your students. A resume project works best with students who are just beginning or are in the early part of their major coursework. These students are often in their third or fourth year of college. They are beginning to take their major courses and may have completed or are about to begin an internship or co-op experience.

A letter and resume project tends to be much less useful in classes where most of the students are beginning their college careers. Writing an effective resume means conceptualizing one's major in terms of a professional community that lies outside the classroom and telling a story that communicates strengths as a potential member of that community. A student taking mostly general education courses (and who may not know what his or her major will be) is usually not ready to think about a professional identity within a professional community.

I have also had less success using a resume project with graduate students, adult learners, or students in their last semester who are actively involved in a job search. Although such students may need help with their resume, they often have so much invested in that resume (they may have used that resume successfully in several job searches) that they resist the rhetorical rethinking and remediation required of a hypermediated resume.

Available Technology

In most cases, access to technology is not a problem in a hypermediated resume project. Most students have or can get access to a quality word processing application and an ink jet printer. Similarly, most students come into class knowing how to use email so the process of creating text-only resumes tends to go smoothly once students make the conceptual leap of removing formatting to create rhetorical emphasis. Only for HTML resumes does technology continue to be an issue.

Mauriello, Pagnucci, and Winner (1999) have written about the difficulties of incorporating Web-based writing assignments into their classes. These difficulties start with the software used to create HTML resumes. Six years ago, when my students first started doing Web-based projects, I taught them how to use code-based HTML programs such as Home Site or BBEdit; it was an enormously time-consuming and intellectually exhausting process. In recent years, however, the quality of Web-authoring software has improved dramatically. Programs such as Macromedia Dreamweaver, Microsoft FrontPage and Adobe Go Live have all but eliminated the need for students to work with the HTML code underlying their Web-based resume. After trying out a variety of these programs, I have standardized the Web-authoring program that comes with Netscape: Netscape Composer. It is simple and straightforward for creating a simple, one-page document like a resume, and any student who has Netscape has Netscape Composer on his or her computer.

Once students have created HTML resumes, they can visit clip-art sites on the Web (such as www.clipart.com) for backgrounds, rules, and other graphical elements, and they can use online tools (such as www.3dtextmaker.com) to create graphical headers.

Of course, creating an HTML resume is one thing. Getting that resume on the Web is another. In the past, using an FTP program to transfer files over the Internet has been a stumbling block. Learning the grammar of transferring files over the Internet was more technology than many students were prepared to learn. In response to this problem, my university instituted a set of Internet services, including a common WINS-based logon process so that students, faculty, and staff can use the same user id and password to log on to computers and access their mail. In addition, each student has been given a ten-megabyte Web folder on an Apache Web server, and the University computer labs are configured so that these shares are mounted as networked volumes when students log on. Anything a student saves to his or her Web share is published on the Web.[8] These changes have eliminated the need to FTP files. Instead of mastering a new program, students publish to the Web using the same grammar of "Save/Save As" that they use to save all of their files.

Even if your institution does not offer a similar "Save As" Web-publishing process, many Web sites offer free Web publishing via simple file upload forms or DAV file sharing. From Web authoring to graphics to Web publishing, appropriate technologies to create HTML resumes are available to anyone with access to the Internet.

Supporting Pedagogical Goals

The final issue to think about when considering a hypermediated resume project is how such an assignment can support your pedagogical goals. You need to reflect on why you are using the resume if you want to integrate the assignment effectively into your course. Here are some of the many different pedagogical goals a hypermediated resume can support.

The Resume and the Writing Process

The resume can be used to foreground the importance of revision. For me, the core experience in learning to write well is revising a text you care about over and over again until it is as good as you can make it. The resume is a form brief enough and a topic important enough to merit repeated revision.

The Resume as Negotiated Social Space

In writing a resume, students get advice from their technical writing teacher, from members of their department/profession, from on-campus career services, from their roommates, and from the companies they are applying to.[9] Students may have already prepared a resume in one or more other classes. The many different voices heard in the process of creating a resume—the resistance, the conversation, and the negotiation—can be used to talk about the social nature of writing as students negotiate a form for their resume in a manner that will satisfy many different interests.

The Resume as Rhetorical and Narrative Form

One of the appealing features of resumes to teachers (and frustrating features for students) is that there is no one correct format. The resume is a most highly rhetoricized genre. There are broad principles of resume writing for students (for example, entry-level people almost always lead with their education), but within those broad principles, the range of possible rhetorical forms is enormous.

A resume is not a neutral display of a student's life history; it is a selective look that should be tailored to each student's strengths as a job candidate. Consequently, the nature of the categories a student includes in a resume, the names a student gives those categories, and the sequence in which those categories appear all depend on the student's unique background. One student may have a particularly strong sequence of course work and course-related projects, another student may have impressive extracurricular activities, another may have co-op or internship experience, while another may have a particularly impressive work history. Each student must decide what story their resume will tell and what sequences of topics and content can tell that story effectively. If you teach the report or the manual as narrative forms, the resume can be used to introduce the idea of narrative in technical writing to the class.

The Resume as Visual Form

The resume is also a very visual form. Content and purpose are shaped by choices in layout and typographic emphasis. Visual issues do not, however, have to be only about aesthetics (making it look pretty) or about usability (making the resume scannable for someone taking ten to twenty seconds to decide whether to toss your resume in the trash). You can also talk about how the visual choices students make in a resume mark them as members of a particular professional community. Thus, the resume can be a starting point for talking about (and getting students to experience) the interplay between the visual, the aesthetic, the cognitive, the political, and the social forces that pull against one another in the shaping of a document.

The Resume and Correctness

The resume can be used to reinforce the importance of proofreading and spelling checking when producing final drafts and to foster an appreciation for the role of correctness in documents written for real readers. I tell the story of a student who was turned down for an internship at a major corporation and the reason that they gave us was that she had misspelled the name of the company in her letter. The rapid transition from early to late drafts in resume writing, the briefness of the form, and the obvious importance of its appearance make the resume a natural place to talk about the difference between reflectively reading early drafts for organizational and conceptual issues and meticulously and obsessively reading late drafts to eliminate error.

The Resume and Overlapping Projects

I have found that most of my students are used to sequential projects in their classes. They work on one project, get a grade, and then work on another project; or they study material, take a test, and study more material. Most professionals, however, must juggle many different projects at once (paying more or less attention to each project as it matures) rather than finishing one project and starting another in a sequential manner. The discontinuous nature of the hypermediated resume assignment (starting a project, going on to other things, and then returning to that project) can be used to introduce the conception of nonsequential writing projects.

These various goals and purposes really just scratch the surface. Technical writing teachers need to be clear about what they value and why they want to include a hypermediated resume in their class if they are to use the assignment effectively.

REMEDIATING THE RESUME

Inevitably, a hypermediated resume assignment like the one described here leads to remediation: "the representation of one media in another" (Bolter and Grusin 1999, 45). Each of the different versions of their resume students create represents and appropriates the forms of the others. Each critiques and informs the other. The ultimate value of a hypermediated resume assignment may well be the reflection and the critique such an assignment encourages.

Reflection should not, however, be limited to students; the hypermediated resume invites teachers to reflect on our practices as well.

Quoting from Marshall McLuhan, Bolter and Grusin (1999) argue that "the 'content' of any medium is always another medium" (45). So, too, the content of any pedagogical activity is always another pedagogy. Innovation in teaching is a form of remediation, a form of representing and appropriating old practices in new. What we learn in teaching the hypermediated resume ultimately emerges out of this remediation, out of the conversations between approaches.

I have argued elsewhere that the history of teaching with technology is a history of remarkably similar patterns (Kalmbach 1997). What teachers did with typewriters in the 1930s and the arguments made for the use of that technology are virtually identical to what teachers did with computers in the classroom in the eighties and the arguments that we made for this technology. From these patterns, I developed the slogan: "Technology changes, pedagogy stays the same." This slogan suggests that the fundamental role of technology in the classroom may well be to create new spaces in which students and teachers can revisit old arguments, old conversations about teaching.

From the perspective of the hypermediated resume, however, we might better argue that teachers use technology not to duplicate but to repurpose and remediate past practice, to revisit and make those practices our own. The versions of the resume I present here will probably change, perhaps dramatically, in the next few years. Technology will provide us with new ways of understanding this genre and its social consequences. No matter; we will create new forms and new practices. As teachers we will repurpose and remediate. That is what we do.

9

USING ROLE-PLAYS TO TEACH TECHNICAL COMMUNICATION

Barry Batorsky
Laura Renick-Butera

INTRODUCTION

The techniques and assignments described in this chapter grow from the authors' own academic interests in constructivist pedagogy, the learning plays of Bertolt Brecht (1968), and Augusto Boal's (1968) Forum Theater. We use role-plays to generate both oral and written assignments, which fulfill two constructivist precepts. First, students learn communications best from authentic problems. Second, learning communication skills is a fully engaging activity: communication is both physical and mental. Specifically, we generate written and oral assignments from role-plays written by students and drawn from their experiences with technical communication problems.

At our school, we teach technical communication to electronics technology students. Our graduates interact with rapidly changing computer and communications technology and must demonstrate high-caliber customer service skills. Traditionally, employers value our graduates because their "hands-on" technical education enables them to "hit the ground running." Upon graduation, however, many students enter company training programs. Increasingly, these programs emphasize customer service and teamwork skills. This emphasis is because, instead of servicing mostly large-scale projects and government contracts, the electronics industry increasingly serves smaller consumers, developing, supplying, and maintaining sophisticated electronic equipment for small businesses and even individual households. Several years ago, an industry representative told one of the authors that companies are almost ready to recruit field technicians from liberal arts programs; companies were willing to provide technical training to applicants who had developed the ability to communicate with customers. In response to these changes in industry, we have developed classroom techniques and assignments to prepare the electronics technician to "come out from behind the machine" and interact with the customer.

Briefly, in our technical communication classes, students role-play personal incidents of failed communication. This role-playing creates the opportunity for a critical distancing of their lived experience. Like Brecht's (1968) estrangement-effects—*V-Effekten* (680)—our role-playing technique gives the audience (and in our case, the student/actor/author) an opportunity to break out of habitual ways of seeing and to observe and engage with the rhetorical choices made by the characters. The intention is for students to develop a sense of control over aspects of their lives that have reified under layers of habit and assumption. Using the techniques of Augusto Boal's (1968) Forum Theater, which developed as a critique of Brechtian theory and practice (139–142), students present and critique their role-plays. The presentation techniques of Forum Theatre empower students to critique their habitual styles of communication. The role-play exercises then become rehearsals for success.

From these role-playing exercises, opportunities for writing naturally emerge. The most usual motivation for writing is somehow to repair or prevent the problems presented in the role-plays. Because they have role-played these problems, students can better visualize the goals of their writing. Those visualizations can then be described, analyzed, and revised, in words as well as in action. Specifying the location for the role-play—on the job or within an organization—assures that issues of organizational and technical communication become the focus of the role-plays. As teachers, we can respond realistically to this writing, posing questions about how the real audience (the audience as presented in the role-play) would react to content, style, and presentation. We become editors positioned between the writers and the audience, which is our preferred teaching position. Thus, our classes become student centered; we become facilitators of learning rather than lecturers, and students become actively and critically engaged in authentic problems of organizational and technical communication.

The following sections describe our use of role-plays, our sources, and some of the broader implications that we think this model has for college pedagogy. In the first section, "The Structure Of The Actor's Work: Creating A Classroom For Role-Plays," we discuss how we prepare students for the work role-plays will demand of them and suggest ways the teacher and class structure can model the dramatic principles and rhetorical discretion inherent in role-plays. In the next section, "The Arsenal: Tools for the Student Creating the Role-Play," we describe a typical sequence for staging and responding to the students' first role-plays of the semester. We refer to the performance concepts of Boal (1968) and Brecht (1968) to clarify the principles and goals of this method of delivering a course in technical communication. In the next section, we describe how role-plays can lead to authentic writing assignments or enrich students' responses to more traditional, textbook-based assignments, including discipline-specific research papers. In a later section, we close with some reflections on how our experience with role-plays has guided our professional lives outside the classroom, including our responses to writing across the curriculum, case-studies initiatives, the pressures of traditional grading, and textbook selection.

THE STRUCTURE OF THE ACTOR'S WORK: CREATING A CLASSROOM FOR ROLE-PLAYS

If we set the scene carefully in the beginning weeks, many classes become self-directed, and we simply guide the work students have already committed to. We begin by introducing fundamental concepts similar to those of most technical communication classes. As we do this, we deliberately model the audience awareness and adaptation we expect from students. The objective of our first class meetings is to get students to articulate their judgments and build their confidence. If we succeed, students articulate authentic situations that guide their work for most of the semester (the subject of the next section), and they commit themselves to finding workable solutions to these situations.

Introducing Rhetorical Analysis

Our technical communication course begins with an introduction to rhetorical analysis. We want to make students aware of the considerable repertoire of rhetorical skills they already have. We lecture briefly about language acquisition and use simple, hands-on demonstrations of the power their language skills give them to organize and communicate information. A simple but powerful example: We give them a list of eight words to memorize in order. As they attempt that, we write the same words on the board so that they form a sentence and ask them to memorize the words again. The sentence is memorized as a unit, almost instantly. They experience, hands-on, one of the most elemental and powerful innate abilities of the human brain. In lecture, we also present some tools for rhetorical analysis, ranging from the basic, traditional author-subject-audience triangle to more complex methods such as Killingsworth and Palmer's (1999) context of production/context of use (3–20).

The level of sophistication of this introduction depends on our sense of the audience we have before us. If the class is anxious to "do something," the introduction is more like a pep talk. More reflective classes do more discussion. We are ready and willing to adapt our presentation, to observe students, and, at any time, to ask the question: "Do you want to do a role-play now?" This willingness to adapt instantiates the basic lesson of our course: An effective communicator responds to the real needs of an audience. From our first meeting, we explicitly model that skill for our classes by eliciting their responses to our plans for the course and our assumptions about communication. This method does not constitute a fundamental change in most teaching styles. Most teaching is as responsive as it is directive, and our approach does not fundamentally alter the mechanics of teaching; instead, it tries to lay bare these fundamentals for students to see and use.

The goal of our introduction is, finally, to prepare students to turn technical data into technical information and to develop from knowers into teachers. Thus, our introductory lecture- demonstrations also have to build students' confidence in their ability to do the work of teaching. Sometimes, technical communication can intimidate even students with a substantial record of success in their technical courses. Technical communication is different from their technical courses both in subject and structure, and we don't try to smooth over the differences. Instead of encountering brand new technical content and a clean slate to write on, students in our courses are made to question some deeply held beliefs and break some old habits. One of these habits is eschewing ambiguity in the search for the one right answer, the one method, for solving all their communication problems. From the first day of class, our work is a matter not of finding the one right answer, but of choosing the best answer from many possible answers. We model for students the need to adapt, and we introduce a range of tools, including our textbooks and other reference material, for informing our judgments and making effective rhetorical choices.

We don't want our introductory classes or our textbook to preempt the students' own decision-making process. We want them to develop confidence by discovering and exercising the rhetorical skills they were born with. We are then able to demand that students test and revise the solutions they suggest to real problems. We try to convince them that human beings are sophisticated rhetoricians long before they are trained technicians.

Integrating Role-Plays

The way we integrate role-plays into our classroom varies from semester to semester and section to section, precisely because the rhetorical situation we face as teachers changes with each set of students. Some semesters, role-plays may serve as a warm-up for a more traditional "genre studies" class. Other semesters, a set of role-plays may bring up issues so complex that we will spend the better part of the semester on them, covering material from textbooks by way of the role-plays and conducting necessary research. In every case, however, by midsemester most classes, whether based on textbook readings or role-plays, have covered the same range of material required by our department's course description. Revision, proofreading, ethics, letter format, memo format, technical descriptions, and so forth emerge from a class centered on role-plays as consistently as they emerge from a syllabus designed around a textbook; but, in a class centered on role-plays, the control over these topics becomes a shared enterprise between teacher and students. Students who emerge from classrooms where role-plays have controlled the syllabus are as well prepared for our departmental writing assessment as those who emerge from classes that more closely follow the structure of the textbook. This preparation is because they have internalized problem-solving strategies that enable them to creatively and confidently address a broad range of technical communication tasks.

A short description will illustrate how our method can structure a semester. One semester, our classes took on the project of revising the labs in their technical courses. This project took the entire semester. It emerged during the first weeks of role-plays. In role-plays based on in-school communication problems, it became clear that students felt they were receiving poor "customer service" in the lab. When they began to explore the systematic causes of this situation, they discovered that the way the labs were written predisposed teachers and lab assistants to offer poor customer service. For instance, cautions and diagrams were not included. When they began to rewrite the labs, they had to consider themselves and their fellow students seriously as an audience. But they also had to consider the larger academic systems within which labs are written and revised. They discovered students in the lab were not their only audience; so were the teachers who would teach (and approve) the changes, the dean who would approve the cost of recopying the labs, and so forth. Each class session began with the question, "Did we get the labs changed?" Answering this question took the entire class session and set the homework for the week. The class agenda and schedule were set by the students' real need to resolve a real problem. They used their communication textbook and their class time as resources to help them effect change, not as ends in themselves. Two products emerged from their work: new labs and a new awareness of their technical classrooms as rhetorically complex environments where students and professors have myriad roles. This authentic writing situation also supported the formal academic requirements for technical communication: students who participated in this project wrote memos, technical descriptions, procedures, and directions; they created charts, diagrams, and visuals; they learned to format documents on the computer; they made oral presentations; and they practiced audience analysis, revision, editing, and proofreading.

The students in this classroom, and in other classes where a role-play lasts a few minutes or a week, have faced the questions of engagement and responsibility crucial to generating and evaluating writing both in the classroom and the workplace. It is easy for students to dismiss rhetorical problems in which they are not fully engaged. Every teacher has worked with students who write merely to complete an assignment and please the teacher. These students can't understand why you are not satisfied when they have done everything you asked—everything, that is, except commit themselves. When presented with a communication problem in a classroom setting, such students confidently ignore it and turn it into something easier to handle, but irrelevant. A problem-solving letter will degenerate into a tirade and end in an inappropriate action, or it will degenerate into a polite, powerless, but structurally perfect "letter of complaint" and end in inaction. However, once students experience the sophistication of their rhetorical powers, they are more prepared to reconsider their first knee-jerk responses to the communication problems we present. For this reason, the problems we present for students' consideration must be fully engaging, both mentally and physically. Such engagement emerges from student-generated role-plays based on situations often ongoing in their lives and on significant events in their working lives. Thus, even if the entire semester does not revolve around role-plays, we find them a useful way to introduce the content, structure, and expectations of our course.

THE ARSENAL: TOOLS FOR THE STUDENT CREATING THE ROLE-PLAY

Once we have set the scene for the actors' work by introducing the fundamentals of the communication situation and establishing student engagement, we turn our attention to generating the actual role-plays that are the meat of the course. We begin with a basic process, which we modify depending on the students and the progress of the semester. Students begin by generating ideas, which we review and categorize to create structure in the course. Students rehearse briefly, and the class is given a topic to consider as they observe the role-play. After each role-play, students discuss the focus topic, and then the players replay the scene, suggesting alternative actions.

Generating "Scripts"

Crucially, students generate the ideas and scripts for the role-plays from their experiences, using a process based on Augusto Boal's (1997) Forum Theater. When students generate, select, and present the situations, they become "spect-actors" rather than spectators (17). This technique empowers them to establish the problems to be addressed and leads them to take responsibility for their communication strategies. This technique also assures an authentic experience of rhetorical problem solving.

It is a relatively simple technique. For the first set of role-plays, we often ask students to describe a time they have seen or experienced a failed attempt to communicate in the workplace. For each role-play, we give the class at most a half hour to produce a script. We leave the definition of "script" wide open. The idea of student-generated, and therefore authentic, scenarios comes directly from Boal's Forum Theater (1968) technique (132). Some students create scripts that simply sketch the outlines of events; others write out complete dialogues. Once they begin to perform their role-plays, the students soon learn that the end result of either script is the same because they must perform without the script.

Especially for the first role-plays, we generate some discussion to prepare students to write their role-plays. A couple of students almost always immediately understand the assignment; thus, as soon as we are done describing the assignment, we ask the class for an idea for a role-play, and we discuss its characteristics. Soon the whole class understands, and everyone can begin to produce a script. Usually, in a class of twenty-five we have found that only one or two have trouble getting a situation after a little thought. Students in our classes are probably older than most, and almost all students have jobs, so results may vary, but we suspect not by much. After all, the sense of what constitutes a rhetorical problem has its roots in an innate human ability to use language. Asking college students to identify a communication problem at work merely taps into the rich and varied experience they already have with language. After everyone has written a script, we collect the scripts and take them home to read and categorize. We look for three categories of circumstances because in the next class we break students into groups of three to work on their role-plays.

Discovering the Curriculum

Discovering these categories requires us to relate the rhetorical lessons we teach to the experiences of the people we are teaching. Like our course introductions, the categories we find vary from class to class, but we have never failed to find a relevant set of categories. For example, one class's scripts were divided into "failed communication between supervisor and employee," "failed communication between employees," and "failed communication between customer and supplier." Another class's scripts divided into "intentional deceptions," "ignorance about the subject of the communication," and "unintentional misunderstandings." (Sometimes we skip the classification step altogether and just go with the flow. At these times, we enjoy the rich and varied scripts and the challenge of on-the-spot analysis and act like a director or a drama critic, identifying the central *agon,* critiquing the plot.)

When we choose to find categories, as we usually do, these categories need to honor the students' authentic experiences and their ways of naming and understanding these experiences. At this point, we exert control over the curriculum. We could choose role-plays we think will lead to technical definitions or technical descriptions, proposals or recommendations. However, imposing categories like these from our textbook or departmental requirements risks diminishing the role-plays' potential for helping students uncover meaning and practice authentic problem-solving. Asking for a role-play that produced a "technical description" situation will not work because situations do not occur to students as instances for the textbook's exegesis; otherwise, they wouldn't need our course. We have learned to trust that discussions of topics like the "extended technical definition" will emerge naturally out of the work on the role-plays. This delay of academic classification helps students discover the issues and strategies of technical writing on their own and thus to adopt them as their own. The relevance of the contents of a textbook has to be discovered, with the teacher serving as facilitator, not lecturer. Indeed, concentrating on basic rhetorical problem solving makes the teaching of genres and types easier and faster because the genre is taught at the moment students are intellectually prepared for it. Once you recognize that an extended definition will solve your problem, the technique becomes as obvious as a hammer: you pick it up and use it.

At the next class meeting, after the scripts have been classified, students begin to perform their role-plays. First, we point out that the role-plays take place in "real time," so that students need to focus those scripts down to their key moments in ways that take no more than about four minutes. We return each student's script with just a number—1, 2, or 3—on it, and students form groups of three. Each group includes the author of a "1" script, a "2" script, and a "3" script, so that each group will—to the extent possible—work with one script from each category we have defined. (We do not tell the students what these categories are.)

The group then reviews each of its scripts, chooses one, and creates a role-play for it, focusing on the crucial actions. For us, this process of focusing is an application of Brecht's (1968) concept of the *gestus,* a social comportment captured in gestures, and Boal's (1968) use of participant-generated action (689). This focusing on the essential action of a scene is a wonderfully complex act of rhetorical analysis. Students accomplish it with ease. The whole process of reviewing, selecting, and "rehearsing" the actions estranges them in ways that make them more available for student interventions. This process can reinforce the students' sense of control and competence. It also prepares students to write more detailed narratives and descriptions. At the end of their "rehearsal" time, each group will have discussed three role-plays, one from each student, and practiced one for presentation. This whole "rehearsal" is remarkably easy and quick.

The first time a class does the exercise, the selection and rehearsal can take about half an hour. After one experience, the role-plays can usually be selected and prepared in less time. Giving more time can even be counterproductive. Students' impromptu responses are most effective at communicating the main issues as soon as the "actors" have a sense of the reality of a scene. Practice here may sometimes dull the edge because students tend to want to soften the conflict. It is often the restless, nonreflective student who produces the most compelling role-play, while the more diligent student sometimes takes fewer risks and produces a two- dimensional representation with a prepared solution. As we said, humans are naturally masterful rhetoricians and, given enough time, students will take any "case" and try to define the rhetorical problem out of existence, especially when it involves written communication. Students who resist role-plays want to wave a wand at the problem; they say, "I wouldn't write; I'd just talk to the guy" or "That's just the way it is; there's nothing I could do about it" and believe it will go away. This ability to "resolve" challenges by redefining the issues is not unrelated to Boal's (1968) concept of "magic." Magic, in Forum Theater, describes proposed solutions that are unrealistic. In Forum Theater, the facilitator, called the Joker, may interrupt a scene by calling out the word "magic" when some proposed resolution seems unreal, such as resolving an oppression by appealing to the oppressor's sense of brotherhood (139–42). The need to overcome this remarkable human ability to sidestep rhetorical challenges is one of the things that brought us to role-plays, and especially to impromptu role-plays, which do not allow students the time or distance they need to bury the key rhetorical issues in diversions and vagaries. In addition, we extend the power of the Joker *mutatis mutandis* to the whole class: that way, even when a student does trivialize his or her role-play, we get a teaching moment. First, the student's attitude is usually the result of an insensitivity (or oversensitivity) to the issues involved. Second, other students can often see the threat and the opportunity that the author is ignoring.

Before we let the students present the first role-play to the class, we usually introduce a focus for the discussion that will follow. For our first set of role-plays, we usually ask students to observe the responsibility each persona in the role-play has for the failure of the communication. Introducing a theme like responsibility just before students see a role-play helps focus the discussion of alternative actions that follows the presentation. If we don't introduce a focusing issue to encourage students to stop and analyze, their innate ability to redefine rhetorical problems sometimes will preempt a discussion of the actual action of the role-play. We like the issue of responsibility because it is central both to the students' work lives and to their comportment in the writing classroom. Raising the issue of responsibility early helps them to find the connection between themselves and writing. For instance, in one role-play, the student had refused to shovel on a work site because he thought it was unsafe. A fellow worker did as he was told and ended up in the hospital. The focus question allowed the discussion of this incident to center on responsibility, not blame: What were the responsibilities of each worker and of their supervisor? A central problem in getting students to recognize the need for rhetorical analysis is teaching them to fix the problem, not the blame. After students are comfortable with the role- play process, we can set a focus for the class discussions at various points in the process: before students begin to script, after we return scripts, or before students rehearse.

Performing the Curriculum

After scripting, rehearsal, and focusing, the actual performance of the first role-plays occurs. Students perform without referring to the written scripts. When we first started using this form of role-playing, we allowed the students to take a copy of the written script with them for a first run-through, but then we had them replay the scene without the script. Students quickly saw that effective role-playing was easier than they thought and could even be fun. On their own, they abandoned the written scripts after rehearsal.

Sometimes, we videotape the role-plays to help resolve disputes about which action caused which reaction. Also, information about body language and physical expression is absorbed in the form of a "dramatic" image, and we have seen what a powerful effect the internalizing of this projected self-image can have on the development of interpersonal skills. We prefer to use a video projector rather than a TV for playbacks. Perhaps because they are small or because the format is so familiar, television screen images do not seem to capture the attention or have the impact of the near life-size image of a student projected onto a screen. We are always amazed at the ability of these images to hold the class's attention for extended periods. We have videotaped several role-plays from one class session (our classes run usually two hours) and held discussions that did not require the use of the film, and yet the students were ready and eager to sit through a replay of all the role-plays at the end of the class. We have also used videotaped role-plays as the basis of exams, asking students to view the tape and then write an analysis of the situation presented. Videotaping is a useful option, but it is not essential; and we wouldn't insist on reviewing a tape before we began the discussion section of our role-play exercises, especially because these discussion may lead back into role-playing as described later.

A discussion of a role-play usually follows immediately after its presentation and can take different forms. The discussions can be short and limited to one aspect of analysis. Often, in the first set of role-plays for a semester, we run through the role-play discussions with just one task: to name the responsibility each persona had in the failure of the communication. We want to reinforce the idea that even the most obvious villain—the boss with a chip on her shoulder or the alcoholic co-worker—to some extent may be enabled by the responses of other actors. On the other hand, the discussion of one role-play could be material for more than one class session. It might open up topics like teamwork, management, or information flow that may require research or reference to the textbook. This dynamic is the same as the one that traditional case-study techniques engage.

Eventually, discussion may lead to alternative ways to envision and enact the role-play. In the style of Boal's (1968) Forum Theater, the students become "spect-actors": not passive observers, but active participants, and therefore active learners. In our discussions, students are able to intervene in a role-play after its first run-through to test out different strategies. For example, in one role-play a student is fired for handling a customer when it wasn't part of his job description. In this situation, the policy conflicted with best practice/policy implementation—an authentic and complex situation. Originally, the student author/actor had not tried to keep his job. Presenting the event as a role-play enabled him to revise his personal history. It was too late to change the outcome, but seeing what he might have done increased his sense of control. He actually considered contacting his former employer, but decided that instead he would accept it as a lesson learned. Thus, during the discussion phase, we employ Forum Theater techniques to allow class members to stop the role-play at critical junctures to introduce other possibilities and reject alternatives as magical. Following Forum Theater practice, we sometimes play each intervention through until we see a real solution or until we come to another intervention that revises the scene or advances the action to move us closer to a resolution of the core problems (Boal 1968, 139–42).

The essential part of the role-play is its function as a rehearsal, estrangement, and deliberation, all which Brecht (1968) and Boal (1968) name as essential to the dramatic process. For example, in one student-generated role-play, this question was posed: How do you respond when a group member isn't pulling his or her weight? Students role-played it two ways. First, they lectured and punished, and the delinquent student-actor naturally (unrehearsed) pulled away and acted worse. Next, they included him, asked for his input, and he responded positively. Rehearsing their response allowed them to respond more deliberately and more effectively. As Brecht observed about his learning plays, this role-playing is a rehearsal for real life.

LISTENING TO WHAT WE HEAR/SEEING WHAT WE LOOK AT: FINDING WRITING WITHIN THE ROLE-PLAYS

Using the process described previously, students generate a wide range of role-plays, practice their interpersonal and self-presentation skills, and discuss thoughtfully many rhetorical issues that arise in technical communication. But students also need to write. Like role-plays themselves, the writing assignments that emerge from them are complex. They invite student investment and revision, place the teacher in the role of facilitator, and teach the principles of technical writing as surely as textbook assignments. These writing assignments, in fact, prepare students for more traditional academic assignments, such as research papers, which we sometimes assign in the second half of the semester. In fact, by midsemester, students are prepared to see their academic writing as role-play, a strategy that gives them the rhetorical tools, sophistication, and confidence to succeed as academic writers.

Writing a Solution

At some point, a group's role-play will generate a writing task, either as a follow-up to the event role-played or as an alternative to the failed strategy of the role-play. This task usually emerges naturally in discussion but can be prompted by questions such as "What needs to happen now?" or "How could this have been avoided?" These questions serve as a springboard for all our writing assignments, which may include letters or memos from one persona to another or to a figure implied by the action. In this way, the writing situation is authentic, visible, and motivated. Students begin to see writing as a natural by-product of life, not as an academic exercise. At the same time the writing assignment remains a common topic for all students in a section, which is practical for our grading purposes and for documenting our adherence to institutional course requirements.

As students draft their writing assignments, we can return to role-playing to test the efficacy of their writing. A follow-up role-play can test the effect of the writing on the intended audience. This role-playing peer review can make clear when a student has tried to solve a complex problem with a sledgehammer. In a particularly effective scenario, one of the students, playing himself, insulted his supervisor's mastery of English. The student then sent the supervisor a memo about the encounter. In a follow-up role-play, the supervisor called the employee into his office after receiving his memo about the encounter. The supervisor first impugned the employee's own mastery of written English and then fired him for insubordination. The class cheered. They love a good payback.

Writing Research

Even if we move to more traditional presentation modes, like academic research projects for a technical course, later in the semester, we find the students' work with role-plays enriches their academic work. Once students understand the communication process within an authentic workplace environment, they are prepared to see the academic genres and classroom learning as authentic. For example, in the first assignments, we use a Total Quality Management customer/supplier model to discuss rhetorical issues (Schmidt 1992, 37). In this model all work relations, both internal and external, are identified as occurring between customers and suppliers. Just as a customer comes into a store to get what he needs to do his job, coworkers within a company act as both customers and suppliers to one another; a key to one's professional life in this model is identifying one's myriad customers, their needs, and the effect one wants to have on them. Later in the semester, we may employ this same model to work with students on a research project assigned from one of their technical courses. We ask students, who are used to seeing themselves as customers of the teacher, to see themselves as suppliers of the research project. We begin with the questions that we often use to focus role-plays: Who are the customers in this situation? What effect do you want to have on them? Using the experience of role-playing, the simplistic answer, "My teacher is the customer" gives way to a more complex discussion: How are the students in your research team also your customers? How are the students in your class customers of your research? How can you make a potential employer the customer of your research? Recently, two of our former students told us that they went to a job fair and told recruiters about their research, which led them to job interviews. It was the role-playing in our class that led them to understand that someone in the real world could be interested in what they had done in an undergraduate research project. They were able to see themselves as suppliers of information to potential employers. That is the lesson we think role-playing teaches well: to become an effective part of an organization, you have to understand the role your communication skills do and can play.

With an improved understanding of research and writing as a commodity that an employer might need, rather than a retelling of old information to prove they have done the required reading, students begin to take more responsibility for their research. In our classes, the purpose of research projects becomes to teach the class, as well as the teacher, something about a technical subject that will be relevant to their lives and work. In this way, with the class as a real audience, the research project becomes a rehearsal for life. As part of the project, students present a proposal to the class before they begin. The class has to approve the topic and suggest what it is they might want to know about the proposed topic. After conducting the approved research, the student must then deliver an oral report to the class and a written report to the professor who assigned the project. This assignment creates two very different audiences. The textbook issue of the complex audience thus arises out of an authentic assignment.

Our expansion of the primary customer from professor to fellow students also helps the class resist the urge to see research as facts lifted from books or the Internet. The whole project is now defined by the need to teach a specific audience specific material and to respond to the real needs of the customers. At the end of the student's final presentation and the question and answer period, the class takes a quiz on the material, designed by the presenter. The success of the class on the test factors into the final project evaluation. This test situation is a more authentic situation than a research paper alone. It has a specific but complex audience (with equal or less knowledge than the writers) and measurable results. It also sensitizes students to their responsibility to an audience and to the extent of their control over both situations and information.

DOUBTS AND CERTAINTIES: THE TEACHER AS LEARNER

The previous sections tried to made clear the value we find for role-plays in students' academic lives, but they have not touched on the value this approach has for our academic and professional lives. Role-plays create a classroom environment that engages us equally as it engages students, allowing us to reflect on the questions we find essential to our lives as teachers and academics. At the same time, they streamline the procedural and administrative work of teaching because they make it clear both to us and to students our function in the classroom. We find our work with role-plays has made us reconsider our position on academic issues, such as writing across the curriculum and case-studies initiatives. Role-plays have also forced us to reflect on some of the essential power dynamics of the classroom embodied in traditional grading and textbooks.

Rethinking Writing Across the Curriculum and Case Studies

For us, use of role-plays developed out of our experience with two movements in the teaching of college writing: writing across the curriculum (WAC) and case studies. When the authors first became colleagues, we worked on developing a WAC program for the college. We began by engaging technical faculty in joint writing projects. Though we succeeded in developing joint writing projects, our commitment to constructivist ideas about student-centered classrooms soon created problems. The assignments from the technical courses were not in tune with a student-centered approach. The research projects were not well adapted to the students' technical interests or research skills. The lab reports were mostly fill-in exercises with a paragraph or two of discussion and conclusions. As described earlier, we adjusted one research project to make it more authentic, but we were and are still dissatisfied with the results. The technical component in these projects remains fully defined by traditional course content. The students are given a list of technical topics to select from, all which are covered in the textbook. Our WAC projects were subordinating the authentic rhetorical needs of students to course content. In response, we initiated our technical projects, such as the semester-long lab revision project described previously. Our frustrations with WAC projects made clear the advantages of the more authentic approach to role-playing that we now use. Role-playing, in turn, influenced our revision of WAC projects.

Subsequent and connected to our work with WAC, which continues, was the opportunity to work on a collegewide case-studies initiative. We collected and developed a set of cases for use in technical and general education courses. In the end, the process of preparing cases for unprepared instructors in existing courses was unsuccessful because the cases we developed looked like the cases that many newer professional/technical writing textbooks provided. These cases, conceived in terms of specific book chapters and discussion points, were not authentic learning experiences for the students. They led students to specific material at specific times, instead of bringing material to students only when discussions require it, which is the traditional method of case-studies courses. In the case-study classroom, the development of the course content should be subordinated to the needs of the case work. This is not what happened in our case-study initiative, for two reasons, we think. First, the institutions, discipline, or self- definition of most technical teachers does not allow them the creativity and close attention to an unfolding class dynamic that being a case facilitator requires. Second, the method used to prepare these cases was inappropriate for our classes. Either they were too sophisticated, requiring too much subject knowledge to be effective, or they had to be stripped down to match the professional knowledge of students. The Harvard Business School case-studies model, which underlies many case-study projects, is probably not a good one for undergraduate students or for students without equivalent professional background and experience.

We were frustrated by WAC and prescriptive case studies because we felt they did not go far enough toward a student-centered pedagogy. Then, at a meeting with representatives of the Field Services Managers Association, we asked what we could do to prepare students for employment and got the response, "Teach them to use role-plays." We combined the role-playing technique with the idea of a student-centered classroom, and the result was the technique we have been describing. The work of Brecht (1968) and Boal (1968, 1997) then gave us the theoretical direction and practical tools for dramatizing our technical writing courses. Role-playing has dramatized the classroom for us. In the process, it has challenged our traditional academic assumptions about assessment, student success, and textbooks.

Rethinking the Power Dynamics of the Classroom

Their initial role-plays unintentionally prepared students and ourselves to interrogate the validity of traditional grading methods in the face of authentic experience. This interrogation is a natural by-product of raising authentic issues of purpose and presentation within a traditional classroom. Students will ask, "If I got the job done, how can you grade me?" Or, "If the other student responded the way I wanted in the role-play where I presented my letter, how are you going to grade me down?" The role-plays help students see the relativity of communication, which in turn complicates grading, with its basis in error rather than achievement. In doing so, role-plays create an opportunity for students to begin to consider their methods for assessing their personal effectiveness.

Perhaps this opportunity is one reason that bad students often succeed at role-plays in ways that force us to redefine the behaviors we categorize as "good" and "bad." The success that low- achieving students experience in role-playing makes us conscious of the artificiality of these labels. Role-plays often help the bad students by validating their active, personal approach and by helping them use it mindfully to get what they want. As teachers of adults, we probably cannot ignore their habits, but we can make them aware of these habits and help them bring them under their control.

Good students often create inappropriate role-plays, tests, and presentations; bad students often create dynamic, engaging, and authentic role-plays, tests, and presentations. For instance, good students will create tests with many multiple-choice questions about specific numbers. Bad students will ask for definitions of basic terms. Good students will "hide" their questions within the presentation; bad students tend to advertise them in neon as the presentation unfolds (writing them on the board before they begin, giving cues such as "This is an important word to remember"). One semester a group of good students complained fiercely because a group of bad students got a better grade. How unfair!

How unheard of! Role-plays disturb students because the usual structure that determines how one earns an "A" isn't there, and they don't know what to do. They have to think about the rules. (In theory, role-plays have given them a tool to do this and to revise how they see rules.)

There seem to be several reasons bad students excel. Bad students help each other more in ways that attempt to leave none behind. Good students often see a locus of control in the material or the professor and define presentations and assignments as "telling knowledge." They see everything in the technical field as talking to someone who knows more than they do and to whom they must prove their command of the material. Role-plays do give these students a language with which to name more diverse audiences, but these students often balk at role-plays and research-for-their-peers because it risks undermining the academic structure in which they have found success and safety. Their presentations tend to be complex and their quizzes too subtle. Bad students, on the other hand, often see the locus of control in their peer groups. What other students think of them is more important than what the teacher thinks of them. This perspective is in some ways actually a more realistic worldview. We believe it leads them to seek success in what their peers can get from them. These students, of course, do recognize the locus of control in the material or professor, but they appear to be so intimidated by the control they perceive in these places that they opt out of the structure to be in control of an alternative, "loser" reality/structure. They are the students who say, "I'll do what I can," instead of "I'll do what I'm supposed to" (which the good students say). This attitude is a way of designing and controlling a situation that often results in effective analysis of our role-play problems.

Rethinking the Textbook

Finally, using role-plays has made us examine and change how we use textbooks in our classrooms. In classes structured around role-plays, we do not use textbooks to plan lessons. Instead, we use them to support students in the work that they have set for themselves. The genre parts of the text (for example, how to write a letter) become reference guides to students after writing assignments emerge from role-plays and discussion. We incorporate the sections on audience, user testing, ethics, and so forth into our daily classroom practice. These sections provide a common vocabulary and ethos with which the class can discuss and understand the rhetorical aspects of their experiences inside and outside the classroom.

We have observed that technical writing textbooks are becoming rhetorically sophisticated in their choice of subject. They discuss more thoroughly issues of audience awareness and ethics. The challenge is to create a way of generating assignments that raise these issues with the same sophistication. Turning our classroom into a more student-centered experience by creating student-directed, problem-solving contexts has helped us to make use of the rhetorical thoroughness of the latest generation of texts. The challenge and opportunity is to capture students' authentic communication experiences. As the field of technical communication changes from a static, codified genre, students' needs have pressured the discipline to reenvision what a textbook should be and what it's capable of doing. Our classroom methods are an attempt to make available to students the possibilities of the new texts and to demonstrate the relevance of rhetorical education.

10

WHO ARE THE USERS?

Media Representations as Audience-Analysis Teaching Tools

Karla Saari Kitalong

If all else fails, read the manual.

Consumer of High-Tech Product

Although usually meant to be humorous, this clichè underscores a truism: Instruction manuals are often unhelpful and difficult to understand. In fact, the quote's bitter tone suggests that customers feel alienated from technological products and that they blame such alienation on flawed instruction manuals. Of course, the technical communicators who write these instruction manuals do not set out to alienate customers; in fact, just the opposite tends to be true—most technical communicators consider themselves to be user advocates.

Nevertheless, many technical manuals are, indeed, unintelligible to their intended readers. This problem affects two kinds of stakeholders. Obviously, unintelligible documentation affects technology consumers, the primary audience for documentation. But it also affects technical communicators, whose reputations and job satisfaction hinge on producing products that users can relate to. Audience analysis is touted as the way to get in touch with the users of technical documents; however, this mainstay of technical communication pedagogy and practice has changed very little in the past thirty years or more, despite all the technological changes that have occurred during that time period. Even a cursory review of the audience-analysis chapters in a selection of recent technical communication textbooks reveals that regardless of the theoretical allegiances of the authors, all textbooks recommend practically the same procedures for analyzing audiences. In most cases, textbook authors recommend a *classification* model like the one outlined by Karen Schriver (1996, 153). Such seeming agreement among textbooks concerning audience analysis implies that technical communicators have settled on the best way to understand readers. As Jan Youga (1989), author of *The Elements of Audience Analysis,* puts it, "When the concept [of audience] is explained to us, we can all nod in agreement at this commonsense notion." However, I suggest in this chapter that the taken-for-granted classification method of audience analysis, while necessary, is not sufficient, especially given recent and continuing changes in both the technological landscape and the users who populate it. As Youga puts it, "to really understand what audience is and how it affects a piece of writing, we need to look at it more closely" (2).

To look more closely at audience analysis in technical communication, I characterize it in terms suggested by J. MacGregor Wise's (1998) concept of the differentiating machine and Bruno Latour's (1993) concepts of purification and hybridity. I propose an alternate or supplemental approach to understanding audiences that blends figural analysis, a method drawn from cinema studies, with what Schriver calls intuition- and feedback-driven audience-analysis methods (1996, 153–154). This alternate method involves regarding as representative users the figures who populate media representations such as advertisements, news reports, and cartoons.

WHY ARE TECHNICAL DOCUMENTS DIFFICULT TO UNDERSTAND?

It's easy to blame technical communicators for inadequate technical documents and for the accompanying alienation that readers experience, as captured in the familiar saying with which I introduced this chapter, "If all else fails, read the manual." A number of authors, probably unintentionally, ratify such blame by recommending that technical communicators address documentation shortcomings by reconsidering how we teach and conduct audience analysis (for instance, Dobrin 1989; Alred, Oliu, and Brusaw 1992; Schriver 1996; Holland, Charrow, and Wright 1988).

Besides blaming readers' alienation on the writers who produce technical documents, one might also logically blame the readers themselves, reasoning that they are responsible for their technological weaknesses. Expressions such as "idiot-proof" and acronyms such as "ESO condition" ("Equipment Smarter than Operator," a term I learned from a student who formerly worked at a home electronics superstore) point unsympathetic fingers at users who do not understand technological devices.[1]

But professional user advocates and interaction designers disagree with both these blame games. The problem with technology, these experts assert, is not that users—or technical communicators—are incompetent, but that technologies are needlessly complex (Norman 1990, 1994; Cooper 1999; Johnson 1998b; Head 1999; Mirel 1993). Users will remain alienated from technology and from technical documents, this line of thinking asserts, until those users have a say in how technological tools are designed. In recent years, a few forward-thinking companies have begun to implement user-centered design (UCD) in which representative users are involved in the design process from the beginning. Other progressive companies perform usability testing, usually late in the product development cycle, in an attempt to intercept and correct major usability problems before the product is released. But neither user-centered design nor usability testing has become widespread enough to prevent end-user alienation, especially because technology use itself has simultaneously intensified. If technical communicators, interface designers, or other user advocates were in control of their companies' product development processes, the transition to a more user-centered product development process might pick up speed. Unfortunately, such control usually rests with professionals like programmers or operations managers, whose priorities are elsewhere (and who coined terms like "idiot-proof" and "ESO"). However, technical communicators can influence segments of the product development process over which they exercise control, including audience analysis, to lessen the impact of customer alienation.

ALIENATION

Just as I don't believe that technical communicators deliberately alienate computer users from technology, I likewise don't believe that computer users have always been alienated in this way. Such alienation appears to be a relatively new phenomenon, fueled by two recent and related technocultural developments. First, there is the sheer proliferation of technology in myriad forms, from cell phones to ATMs to computers.[2] Everyday lives in Western societies are increasingly organized and overdetermined by technology: "Like the purloined letter, technology is 'there'" in plain view, yet it "cannot be located in any one place" (Ormiston 1990, 102). Not so long ago, the typical computer user was a professional computer scientist, engineer, technical researcher, or advanced technical student. In J. Macgregor Wise's (1998) terms, a *differentiating machine* was in place that successfully organized—or differentiated—technological artifacts and their users, so as to create easily distinguished categories of technology expertise (67). Today, on the other hand, ordinary citizens participate alongside technological specialists in the effects of technological growth and overdetermination.

> Traditionally, the more complex a mechanical device was, the more highly trained its operators were. Big machinery was always locked away and operated by trained professionals in white lab coats. The information age changed everything, and we now expect amateurs to manage technology far more complex than our parents ever faced. (Cooper 1999, 34)

Even as both work and leisure activities compel us to operate, manage, and—we hope—also understand an unprecedented array of complex technological systems, information about technology comes to us, unbidden, from a variety of sources. Given all of this complexity, perhaps the instructional manual is the least influential source of technology information for most people. Although it is the mainstay of technical communication work, we have already established that the general public is granted license—by each other and arguably also by the media—to ignore instructional manuals, to read them only "if all else fails."

The previously mentioned proliferation of technology has led to a corresponding proliferation of users, who are now more fully diversified than ever before in terms of the traditional audience-analysis categories of educational background, profession, age, gender, race, and economic status (Cooper 1999). With the increase in technology use and the accompanying increase in users, then, a differentiating machine that functioned adequately in the past no longer creates useful categories of technological expertise (cf. Rubin 1994, 5–6).

Despite recent changes in technology habits, technical communicators still rely upon a simplified binary differentiating machine to classify audiences. First, readers are usually classified with respect to their tool proficiency, such that the tool separates novice from expert. This audience-analysis system prevails in textbooks aimed at technical communication majors (for example, Alred, Brusaw, and Oliu 1992), as well as in textbooks designed for technical communication service courses (such as Burnett 1994; Lay et al. 1995).

The differentiating machine also places writers in contention with their audiences (Youga 1989, 39). Like the binary that attempts to distinguish novices from experts, the writer/audience binary emphasizes difference and opposition. Technical communication scholarship and pedagogy often suggest a user-advocacy role for technical communicators (see, for example, Redish 1993; Dobrin 1989; Cilenger 1992; Wells 1986). But I submit that the contentious writer/audience relationship outlined previously discourages technical communicators' user- advocacy role; thus, despite their best intentions, technical communicators may contribute to users' feelings of incompetence and alienation.

HYBRIDITY

For an alternative to habitual binary constructions, we might turn to Bruno Latour's (1993) work in *We Have Never Been Modern.* Latour posits hybrids, "mixtures of nature and culture," as a counterpoint to so called "purified" categories such as binary oppositions, which are mutually exclusive; that is, they do not overlap and cannot be conflated (30). Latour argues that the modernist paradigm disavows the simultaneous existence of purified and hybrid categories. Although he demonstrates that hybrids exist in a modernist world, Latour claims that they are not acknowledged. They are "invisible, unthinkable, unrepresentable" (34).

Novice and expert are purified binary categories.[3] I submit that the privileging of purified categories limits current audience-analysis pedagogy (and may also limit practice, although additional research is required to verify this hypothesis). Like Latour, I believe in hybrids. In fact, I regard today's computer users as hybrids. Changes in the distribution and use of technology—changes that have led to the emergence of the hybrid user—suggest a need for technical communication teachers to develop an audience-analysis pedagogy that *disrupts* the binary differentiating machine's work, that *contaminates* the purified categories of novice and expert. One way to do this is to introduce students to media representations of computer users through a figural-analysis methodology, which involves close examination and speculation about the characteristics and motivations of the figures depicted in media representations.

Studying media representations in the technical communication classroom has several benefits. For one thing, media representations can suggest metaphors with which to frame instruction sets. If groups of technical communication students are assigned to survey technological metaphors used by journalists or advertisers, the resulting awareness of how technologies are described in the media can help them understand how their readers are taught to conceptualize technology. Moreover, such a focused attention on technological metaphors can suggest strategies for countering unproductive metaphors as well as opportunities for building on sustainable metaphors. If the media use a particular term or concept to describe a technological process, for example, perhaps that term should be included in a document's index as a cross- reference to the term that the product employs. How do popular magazines refer to Web navigation strategies? If "surfing," "browsing," and "searching" are all popular terms, then all should be indexed. Similarly, hardware technical writers who observe that a competitor's ads promise an interesting feature that is perhaps not fully developed may be better prepared to explain the drawbacks of such a feature to their intended readers.

Moreover, as the figural-analysis methodology outlined in this chapter suggests, studying the figures presented in media representations of technology can give technical communicators a sense of a broader cultural view of the technology user. If popular culture represents society thinking about itself (Asimow 2000, n. 90), then images of technology users in popular culture represent what society thinks those users are (or should be) like. Asimow (2002) authorizes the study of popular culture in professional contexts: "The fact that works of popular culture tend to reflect (at least in distorted form) popular attitudes, misconceptions, and myths is itself important and justifies the study of these works as a barometer of public opinion" (550).

Of course, technical communicators are not accustomed to seeing relationships between their work and media representations. But in fact, the media engage in cultural pedagogy: they teach us, in multiple and diverse ways, how we are to understand and interpret culture and its contexts, including the contexts in which technical communication takes place. When I teach figural analysis in my technical communication classes, I begin with two cartoons that appeared about eighteen months apart in two different publications (*Newsweek,* February 27, 1995, 21, and the *Chronicle of Higher Education,*, September 20, 1996). Both cartoons depicted white, middle-aged, middle-class men struggling to use their computers. In the *Newsweek* cartoon, the man is working on his home computer; he complains to his wife that he can't figure out how to install a piece of software and then asks, "What's that neighbor kid's name again?" In the *Chronicle* cartoon, a curmudgeonly older man—presumably a professor—works at his office computer with a younger female colleague, who explains mouse operation in patronizing baby talk.

In both these instances, the male figures bear marks of success, such as the ability to make major purchases and work in well-appointed offices. But when a computer comes into the picture, the men's status as insiders is called into question. In both cases, the men are forced to learn *prefiguratively*—from a younger, less-experienced person, instead of in the more traditional *postfigurative* manner, from an older person (Mead 1970, 1).

On the surface, the cartoons can be seen simply as a commentary on the ineptitude so many otherwise competent adults feel when confronted with even the most run-of-the-mill computing tasks. For technical communicators, however, they provide anecdotal evidence of an inversion of the classic novice/expert binary. Analysis of the cartoons, including figural analysis, suggests that a purified novice/expert binary is still highly valued in Western technoculture while acknowledging that such a binary no longer functions as before. Thus, when technical communicators attempt to define expertise, they must acknowledge that so-called experts may have uneasy relationships with technology. Moreover, elders are expected to guide, protect, and educate younger people; however, especially in the *Newsweek* cartoon, age no longer guarantees status: in fact, it may hinder the achievement of certain kinds of status. Scrutinized through a technological lens, both cartoon computer users are found wanting, despite the status they have obviously achieved in other areas. But the prevailing educational model—Mead's postfigurative model in which adults teach and children learn—does not accommodate their learning needs.

Each cartoon character represents what I would call a hybrid user of computers. But the media, relying on purification, depict them as novices and denigrate their limited technological expertise. Similarly, technical communicators faced with writing manuals for these men, and reliant upon purified categories of audience, would also characterize them as novices on the basis of their lack of computer knowledge. However, addressing them as novices denies their achieved status and life experiences that originate outside the high-tech realm, in much the same way as does speaking in baby talk to a university professor. The cartoons offer two choices to adults struggling with technology: either they can suffer experts' patronizing attitudes, or they can step aside and permit experts to do the complex tasks for them. Neither approach suffices; the boundaries between novice and expert are fluid and shifting.

A figural analysis may also be conducted on news representations. By examining news reports, wherein words and illustrations work together, we can see that expertise with technology is again inscribed quite differently from a perceived lack of expertise. For example, the August 8, 1994, issue of *Newsweek,* which featured a Woodstock cover story, also included an article titled "The Birth of the Internet." The article explains that in 1969, while others were enjoying Woodstock, a "small group of computer scientists" was busily inventing the Internet, thereby "changing the future of computing" (Kantrowitz and Rogers 1994, 56–58).

Most of the people who developed the Internet in the 1960s are relatively unknown to us today; a few of the names are recognizable to computer industry insiders, but others mentioned in this article have faded into obscurity. These Internet pioneers had to learn from each other, in Mead's *cofigurative* mode, because there were no elders to lead them in their groundbreaking quest. The cofigurative, or apprenticeship, learning model, with its accompanying disdain for hierarchies, is commonplace throughout the computer industry (Levy 1984) and resonates to this day.

The 1960s Internet pioneers have led lives of relative obscurity. Given the fame and fortune enjoyed by today's Internet developers, one might expect some resentment to be expressed by the 1960s experts, but, as team member Robert Kahn asserts, he "doesn't like to dwell on the past." Moreover, the optimistic idealism of the men quoted in this article makes today's Internet heroes' work seem even more crassly commercial. Kahn, for example, optimistically privileges the myth of progress, as this closing quote from the *Newsweek* article attests: "Those were very exciting days, but there are new frontiers in every direction I can look. . . . A quarter century later, the future still looks bright" (Kantrowitz and Rogers 1994, 58).

This bright future, the foundation of which was laid by the 1960s Internet gurus, is now enjoyed by the experts who have more recently made their mark on the commercial Internet, and the latter-day experts didn't wait thirty years for their achievements to be recognized. But, as a figural analysis of the Internet pioneers illustrates, the experiences of the earlier experts differed greatly (and continue to differ) from the prominence and wealth enjoyed by today's well-known experts. For instance, the most famous of today's Internet gurus, Microsoft's Bill

Gates, routinely appears on the covers of *Time* and *Newsweek*. This exposure is perhaps not surprising; he is, after all, among the world's richest men and his company arguably has set the stage for our turn-of-the-century computing environment.[4] But other technology figures also attract popular media attention.

Netscape developer Mark Andreessen was featured on the cover of *Time* magazine (February 19, 1996) and touted as one of a group of "Golden Geeks." This hero of the cyber-revolution amassed, almost overnight, a wealth of $58 million. *Time*'s cover depicted the barefoot, twenty-four-year-old Andreessen wearing a crown and seated on a throne and responded to readers' interest, not only in Andreessen's and the other Golden Geeks' technological achievements but also in their private lives and their cultural influence: "They invent. They start companies. And the stock market has made them *instantaires*. Who are they? How do they live? And what do they mean for America's future?" Andreessen's overnight wealth is democratic, the article implied, because anyone with comparable intelligence, luck, and timing could achieve a similar status. The myth of progress resonates throughout the article, yet luck seems to play a role as well. These "instantaires" are not just ordinary folks who have achieved the American Dream; they are computer geniuses blessed with an incredible sense of timing. Barefoot, baby-faced Andreessen remains decidedly down-to-earth and modest despite his success.

On the one hand, then, in the mid-1990s, both the Clinton administration and the media were promoting the Internet as a culture-changing technology. Excitement surrounding the commercial prospects of the Internet intimated that no particular expertise would be required to take advantage of easy, in-home access to information, commerce, politics, and entertainment and that unprecedented economic growth would follow the acquisition of new global markets and the invention of innovative communication products and services.

On the other hand, however, in concurrent ads, magazine covers, and news and feature articles, the media presented the people who understood these new technologies as endowed with special powers not available to the average adult. This vision of "golden geeks" suggests a binary differentiating machine at work in a polarized Internet environment. Just as attention to Latour's work suggests, technology expertise retained a special prominence, with "golden geeks" and "instantaires" as the poster children of this exalted and purified category. Because hybrid computer users are invisible to a culture predisposed to see purified categories, then, hybrids' wide-ranging, diverse, and idiosyncratic attributes, skills, aptitudes, and needs are unrepresentable as such in the media.

REPRESENTING HYBRID TECHNOLOGY USERS

As Latour (1993) suggests, hybridity may be masked by the expectation that there are only purified categories of expertise—namely novice and expert. But if we know to watch for them, we can find media representations of hybrid computing expertise. In fact, hybrids are present in two advertisements that appeared in *Web Week* magazine in August 1995. *Web Week* is a Web developers' magazine, aimed at experienced computer users. The center spread of that issue featured an advertisement for the Apple Internet Server Solution. Four experienced professionals of varying ages, races, and genders—marketing manager Lawson Clarke, biology department chair Lisa Honea, yarn shop owner Debbie Heick, and freelance artist Joe Rosales—are pictured in and addressed by the ad.

The advertisement's headline poses a question and answers itself with another question: "Looking for a compelling reason to set up an Internet Web site? How about the fact that you don't have to be a propellerhead anymore in order to actually do it?" Immediately, then, the ad sets up the binary between expert ("propellerhead") and novice (you, the reader). The ineptitude of novices is further alluded to in the ad copy, which begins, "It's called the World Wide Web (WWW). But that doesn't mean that you have to get tangled up in it." In language that alternates between technical jargon and comforting, respectful reassurance, the advertisement explains that the Apple Internet Server "represents the easiest, most affordable way for people to make their information widely accessible on the Internet." Expecting that readers will identify with one of the figures pictured in the ad, the copywriter allays some of their concerns: "Virtually anybody can now create a WWW site" that is "full of hyperlinked text, graphics, video and sound," and that site can be "up and running in minutes at less than half the cost (not to mention the headaches) of a typical UNIX-based server." As we keep in mind that this ad predated free-access Web sites like Geocities and Angelfire and WYSIWYG HTML editors like Dreamweaver and Front Page, a final statement in the ad perhaps reveals the company's attitude toward its target audience: "The sample Home Pages can even be customized," the ad copy reassures us—as though customizable home pages, even in the early days of the Web, were unusual.

Each of the portrayed figures is a hybrid—an expert in his or her own field of art, science, marketing, or crafts, but inexperienced with computers. However, the ad ends up ignoring their hybridity and addressing them as novices based on their technology expertise alone. This is especially evident in that last sentence, but can be seen elsewhere in the ad as well. For example, the costs of mounting a Web site are mentioned, but no hard data are provided. Similarly, the ad mentions, but does not define, the volume of site traffic that can be considered "heavy." In short, the advertisement seems to expect that the reader will uncritically (perhaps naively) accept its claims—that is, that he or she will adopt a novice's mentality.

In contrast, the intended reader of a second two-page ad in the same issue of *Web Week* is addressed as a technological expert in charge of an existing company Web presence. Again, the ad copy makes a number of claims about the capability of the server being promoted—in this case, a Silicon Graphics UNIX-based Web FORCE server. The ad suggests that Web managers must build speedy Web servers to attract and retain customers. If potential customers must wait to enter your site, the ad copy admonishes, "they'll probably move on," which means lost customers and revenue.

The ad appeals to a different kind of hybrid: a system administrator or other technical professional who has assumed Web design and maintenance responsibilities. This techie is the person user advocates would least like to see running a company's Web presence. Lacking in user sensitivity, the techie interpellated by this ad is concerned with maximizing the technical capabilities of the company's Web site. The ad appeals to this reader by mentioning special customer analysis and tracking tools that can enhance a company's efficiency and competitive edge. "These days, it's not enough to set up a home page and wait and see what happens," the ad notes; in addition, you must build a "valuable database of who's visiting and what they are doing" so that you can "generate Web pages on the fly" to meet the specific needs of individual customers or particular categories of customers. The "unrivaled throughput and scalability" of Web FORCE systems "give your creative teams the tools they need" to make your site "attract and retain a crowd."

The image that accompanies the Silicon Graphics ad is particularly startling: it portrays site visitors as a flock of sheep trying to fit through a narrow bridge. By assuming that techies look down on less-experienced users, the ad gets away with referring to customers as sheep.[5] In short, the ad can be interpreted as saying that visitors to a business's Web site, like sheep, will always follow the flock. With text written in a coconspiratorial voice, the ad invites the techie who doubles as a Web master to regard potential customers as sheeplike; furthermore, these customers, although they apparently cannot make a decision for themselves, must be cultivated and flattered by high-speed connectivity and by compelling and individually responsive content, so that they will return to the site and ultimately purchase the product or service being marketed there. In short, we are left with the unsettling impression that Silicon Graphics regards the end users of commercial Web sites as passive consumers easily swayed by technological wizardry.

A PEDAGOGY FOR ADDRESSING HYBRID USERS

The media representations featured in this chapter were gathered over a four-year period from 1993 to 1997, as the Internet was emerging as a commercial force. Because I have worked with them for a long time, my interpretations are fairly detailed and may sound definitive, perhaps suggesting that meaningfully incorporating media representations into technical communication classes requires sustained attention and perhaps even some special semiotic expertise. However, if you pay attention to the content of your daily media dose, whether it consists of newspapers, magazines, television, radio, or billboards, you'll find more examples than you can use. Take Apple's iMac campaign, for example. To be sure, some of the ads in that campaign focus on hardware attributes (like "The New iMac," which appeared on the back cover of *Time*'s April 6, 2002, issue). However, another thread of that campaign features emails allegedly sent to Apple by average people who recently switched from a PC to the iMac. In the June 17, 2002, issue of *Time,* Mark Frauenfelder, a freelance writer and illustrator, writes that he switched because "I wanted a better computing experience than I had with my PC." Mark, pictured on the left-hand page of a two-page spread, at first resisted switching platforms because he thought it would be too much trouble. "I thought, why make the leap? It's like being stuck in a bad relationship: It works on some level, so you don't want to make the effort to change." In the end, though, switching was no trouble at all, and he writes, "I'm GLAD I switched." The iMac, he claims, can do everything he needs it to do—except make coffee in the morning.

Mark is portrayed as a hybrid; although he's tied into computers for his work, we get a picture of him as an individual. We can relate to his comments about bad relationships and the body's need for coffee. He's not rich or glamorous—in fact, his picture shows a downright nerdy guy who, we might conclude, was photographed before he had brewed his morning coffee. Another figure in the iMac ad campaign is Aaron Adams, who says he works as a PC LAN administrator. In his email, he writes, "At work, I deal with PCs all day long and I can say without exaggeration that keeping those Windows machines running is a constant struggle." This ad, which appeared in *Time*'s July 2, 2002, issue, demystifies the work of the system administrator—the definitive techie—by revealing that Aaron regards his work as "fighting with computers." He doesn't mind doing so "on my employer's time" but wants an easier, friendlier computer to play with at night.

Buried in Apple's iMac campaign is an emphasis on cross-platform compatibility and ease of use. By featuring users' difficulties with the complexity of computers, the campaign suggests that even people we regard as experts struggle with computers and maintains that the iMac computer solves user frustration because it "just works" without a fight. One might say that the iMac ad campaign has incorporated hybrid users—people not defined solely by their technology expertise—but it does not yet displace purified categories of novice and expert.

IBM's countercampaign bears out this observation. Structurally similar to the iMac campaign, in that a typical user is pictured on one page with information about the circumstances of that person's use on the facing page, IBM's series of ads features the ThinkPad, a laptop that is the choice of "some of the world's most successful people," including Charles Nolan, a designer for Anne Klein. Nolan is pictured conferring with a colleague at a point "halfway through the fall/winter line." While the IBM ad calls upon us to emulate IBM ThinkPad users who have achieved success in their fields, the iMac ad incorporates the "authentic" user voice of regular working people from a variety of disciplines who have abandoned their PCs in favor of the more usable and flexible iMac. Although polarized novice/expert categories are not absent, in this genre of ads, we are introduced to hybrid users.

In proposing ads such as these as tools for audience analysis in upper-division technical communication courses, I might ask students to answer a rather simple question: "Who does the ad want us to be?" They articulate excellent and detailed descriptions of Aaron and Mark and Charles—the figures in the ads become very real to them. Nonetheless, in more general discussions of audience, some students proclaim that there's no excuse for user frustration; technology is just part of what we do as members of Western technoculture, they assert, so users need to get over their frustrations and get with the program. Thus, media representations of frustrated technology users are countered for many of them by their experience as successful technology users—graduating seniors in a technical communication program. The digital divide doesn't affect them much; in fact, every student in the class owns a late-model computer with Internet access, and class is held in a computer lab equipped with state-of-the-art computers and recent versions of key software. Although they hear about the digital divide—the "haves" polarized against the "have-nots"—these students don't have direct experience with that divide. Moreover, despite several years of classroom talk about audience analysis and user advocacy, some of them still exhibit a "blame-the-user" mentality. Media representations can be cited as the source—or at least as a reinforcement—for such a mentality. A case in point is Robert J. Samuelson's editorial, "Debunking the Digital Divide" (*Newsweek,* March 25, 2002, 37). In the middle of a technology-oriented issue of *Newsweek* that featured a cover story entitled, "Silicon Valley Bytes Back," Samuelson concludes his editorial as follows:

> The "digital divide" suggested a simple solution (computers) for a complex problem (poverty). With more computer access, the poor could escape their lot. But computers never were the source of anyone's poverty, and as for escaping, what people do for themselves matters more than what technology can do for them. (37)

It's hard to argue with Samuelson's contention that we are being overly simplistic if we assume that the application of computers will solve all the problems that the "digital divide" encapsulates. But isn't Samuelson also oversimplifying when he denies any connection between computers and poverty? Samuelson's essay highlights problems with our typical audience-analysis methods in ways that students and teachers should reflect upon in the context of the technical communication classroom.

In this chapter, I've suggested some ways of thinking about media representations in light of technical communication and outlined a few ideas for including media representations in the technical communication classroom—as aids to audience analysis and as sources of prevailing technological metaphors. Other innovative approaches to integrating cultural representations into technical communication pedagogy remain to be unveiled.

11

WHAT'S UP, DOC?

Approaching Medicine as a Cultural Institution in the Technical Communication Classroom by Studying the Discourses of Standard and Alternative Cancer Treatments

Michael J. Zerbe

INTRODUCTION

Scientific and technical communication has, over the past few years, been witness to a call for research more culturally informed (see, for example, Longo 1998 and Herndl 1993b). This type of research takes into account—and indeed regards as primary—the notion that science is, more than anything else, a powerful cultural institution that has profound and real effects on individuals. Indeed, Herndl (1993a), for one, states that "cultural studies assumes that no undertaking, including science, is autonomous and that any discourse is inherently ideological." Discussing a branch of biology as a specific example of a scientific cultural institution, he adds, "cultural critics might ask how evolutionary biology participates in the whole political and social process of organizing life and legitimizing knowledge and power" (66). In addition, this research recognizes the cultural institution of science operates discursively, thus complicating largely service course–oriented notions of scientific and technical communication as being driven by documents and Web sites analyzed or produced in isolation. Because culturally informed research recognizes texts as a manifestation of cultural institutions and because we as scientific and technical communicators are by and large keenly aware of this connection, we are both well qualified and well positioned to study the discourse of science and to use those analyses to comment on how science functions in the society at large. In fact, we are even ethically obligated to do so, given that science has such a significant impact on our lives and those of students.

Although some attention has been given to the need for and conduct of culturally informed research in institutions and organizations in which scientific and technical communication takes place, less work has been done on its pedagogical counterpart.[1] However, although she discusses research methodology rather than pedagogy, Longo (1998) has provided some insight into how such a pedagogy might be modeled: she states that researchers in technical communication need to use cultural studies research strategies "to illuminate technical writing issues . . . such as effects of institutional relationships and expanded notions of culture . . . as well as the cultural power of knowledge" (54). For us as instructors of technical and scientific communication to fulfill our missions of highlighting the importance and the effects of scientific discourse and teaching students how to analyze and produce this discourse effectively, we must use ideas such as Herndl's and Longo's and develop innovative pedagogies to ensure that students understand that science is—like education, government, religion, the market, and the media, as our colleagues who teach in first-year writing programs have shown us—a cultural construct that impacts students (and us) significantly and repeatedly. Science reflects and reinforces cultural biases, and it operates on the basis of negotiation and persuasion within specific ideological contexts. Students need to understand the wherefore and the why of these biases and contexts so that they can work with and within them as informed participants, be that as professional scientific and technical communicators or as private citizens. Armed with such an understanding, it is my hope that students can *read* science-related texts with a clearer awareness of what is going on between the lines, figures, tables, and other elements and *produce* science-related texts that stand an improved likelihood of accomplishing their rhetorical objectives.

In this chapter, I demonstrate how the discourse of medicine, a branch of the cultural institution of science that not only profoundly shapes individual consciousness but also may soon, as a result of the Human Genome Project, actually *create* it as well, can be used in introductory scientific and technical communication classes, as well as medical writing courses, to help students understand that the cultural institution of medicine operates with specific (but not necessarily always consistent) economic, political, and social agendas that help to form all our subjectivities. I accomplish this by first demonstrating how I would attempt to convince students that medicine is a cultural institution that operates discursively and that has an impact on identity. I then explain how I have asked my undergraduate students, using a set of four questions

I derived from work in cultural studies and in rhetoric and technical communication, to characterize the cultural institution of medicine as a result of their interaction with discourses on standard and alternative cancer treatments. My goal in this research was to determine whether the characterizations differ on the basis of the type of discourse that the students engaged and, if so, why these differences occurred.[2]

Whether a difference appears or not, the more complex understanding of medicine that results from the awareness gained by comparing standard and alternative cancer discourses will, I believe, enable students to take more ownership of their health and that of others close to them, to stand a better chance of succeeding professionally as writers, editors, and designers because they understand more broadly the purpose and potential implications of their work and to effect meaningful change in the cultural institution of medicine as a whole by, for example, critiquing and improving existing genres and suggesting and developing new ones when the rhetorical situation demands it. Whether producing or analyzing texts, the ability to engage medicine more meaningfully at both a personal and an institutional level as a result of cultural-context awareness would, in turn, make possible the ability to engage the broader institution of science more successfully (for instance, to ask questions such as "Who paid for this research?" and "Who benefits from its conclusions?"). Indeed, in our life experience, many profound legislative, judicial, and corporate decisions are made on the basis of scientific research, ranging from gauging the effectiveness of college entrance exams such as the SAT to diagnosing a child with attention deficit disorder. Because Western culture is so utterly dependent on science, students must possess a critical consciousness of these institutions to participate fully in the workings of society.

SHOW ME THE THEORY: CONVINCING STUDENTS THAT MEDICINE OPERATES AS A DISCURSIVELY GOVERNED CULTURAL INSTITUTION

Medical discourse affects students directly. College students are often asked, in the form of surveys, about health concerns considered important to them: alcohol and drug use, physical health, psychological health, and sexual attitudes (Sax 1997). In addition, college health centers and wellness programs use the Web extensively to provide health-related information and promote prevention practices (see Fulop and Varzandeh 1996, for example). That medicine is a powerful cultural institution there can be little doubt, and I would try to persuade students to agree by introducing some cultural studies terms and figures as follows. By cultural institution, I mean that medicine can be understood, for example, as an Althusserian ideological state apparatus. Althusser (1971) defines such entities as "a certain number of realities which present themselves to the immediate observer in the form of distinct and specialized institutions" (143). Althusser does not include medicine in his list of ideological state apparatuses, but it certainly meets the criteria he establishes for them. Indeed, medicine produces a product—health—and, simultaneously, reproduces the means and conditions of its production (128), in that, even as it finds a cure or treatment for some condition or disease, another previously unknown ailment has likely been identified to challenge researchers and clinicians anew.[3]

As do all cultural institutions, medicine affects individuals (see Berlin 1996 and George and Shoos 1992 for a discussion about how, by way of signifying practices,[4] cultural institutions such as media and the market achieve this impact). That is, it significantly influences subjectivity, and, of course, in the classroom I would emphasize student subjectivity. Perhaps no better way can be found to demonstrate this subjectivity than by briefly examining with students the Human Genome Project, which is the largest scientific undertaking of our time. Similar to those who worked on the development of the atomic bomb in the 1940s and early 1950s or on the Apollo missions of the 1960s, scientists working on the Human Genome Project often see themselves as part of a grand, noble endeavor that will inalterably change the way people understand science and medicine. Myers (1990) has demonstrated how the gene has become a cultural icon that people look to as a source of, among other things, individual identity. Because icons can be understood as "a sign in which there is a relationship of resemblance between signifier and signified" (Myers 53), one can argue that we are reaching the point in Western society in which we are defined by our DNA (see such recent films as *Gattaca*—a film relatively popular with students, I've found—for examples of how this idea plays out in popular culture). DNA dictates physical characteristics, such as hair and eye color and genetic susceptibility to disease, and may, according to some, substantially contribute to personality traits as well. Thus, the ability to manipulate DNA at will has the potential to profoundly shape and even create identity, starting with selecting a gender and moving on from there.

In addition to impacting individuals directly, cultural institutions operate discursively, and the cultural institution of medicine is no exception. Within technical and scientific communication, Lutz and Fuller (1998) have focused specifically on the discursive operations of the cultural institutions of "orthodox" and alternative medicine; they explore how the discourses of alternative and complementary medicine work "through language to gain credibility against, through, and around" the more stable and entrenched institution of orthodox medicine. I would introduce students to these two distinct institutions of medicine. On a more practical level, I would also share with students research such as Koski's (1997), which states that between twenty-five and forty percent of the stories in daily newspapers are health-related. Of even greater importance for students, of course, is the Internet, which is replete with health-related sites. Even in 1997, the popular search engine site Yahoo! listed some 15,000 health-related sites; as of May 2002, that number had risen to over 21,500.

I would ask students to consider that medical information is often couched in scientific terms and is thus able to "partake of the cultural power residing in scientific knowledge" (Longo 1998, 59). Indeed, even Foucault (1977b) contends, "'Truth' is centered on the form of scientific discourse and the institutions which produce it" (73). Because it is often obtained during a time of crisis, many people want to believe that the medical information they procure is true and reliable; when this information is presented within a scientific context, it is more likely not to be questioned critically. Such trust can create problems that can be illustrated for students. The authority with which medical information is vested contributed a great deal, for example, to the ongoing controversy over mammography, which pits a National Cancer Institute–sponsored panel of oncologists against radiologists in a bitter dispute over the necessity of mammography for women in their forties. Essentially, the oncologists found that there was no proven long-term survival benefit gained by mammography screenings of women between the ages of forty and fifty. In fact, the oncologists added that an annual mammography for these women may actually have drawbacks.[5] The oncologists recommended that women begin mammography at age fifty. The radiologists, who, it must be pointed out, benefited significantly from providing annual mammography to women forty years of age and over, fiercely contested this conclusion and pointed to their studies, which demonstrated that women who started annual mammography screenings at age forty were much more likely to detect cancer early, when it is most treatable. Especially exasperating to followers of this controversy was the fact that both of the groups involved in the mammography controversy had mountains of "scientific" evidence to back up their claims. I would ask students whom they would believe. Indeed, many people did not have a legitimate mechanism to evaluate the competing conclusions. As a result, many women—not to mention their physicians—were unsure whom to believe: these competing discourses had the all-too-real effect of causing confusion and anxiety over an extremely important and personal medical decision.

AN APPLICATION: STANDARD AND ALTERNATIVE CANCER DISCOURSES

I asked students in my undergraduate, 200-level technical and scientific editing course, who major in technical and scientific communication, English, speech communication, media arts and design, or chemistry, to examine discourses representative of standard and alternative or complementary cancer treatments as part of a discussion of editing Web sites for development of ethos. I consider discourse about a standard cancer treatment to be representative of the cultural institution of Western medicine, while I consider discourse about a nontraditional (at least from a Western perspective) cancer treatment to be representative of the cultural institution of alternative and complementary medicine.[6] For an example of standard cancer treatment, I chose Web-based lung cancer treatment information provided via the National Cancer Institute's (NCI's) CancerNet.[7] Part of the federally funded National Institutes of Health in Bethesda, Maryland, the NCI is regarded as the world leader in conducting and supporting cancer research, treating cancer, and supplying information about cancer to patients, their families, the general public, and health professionals. This lung cancer treatment information on the NCI Web site displays characteristics typical of traditional Western medical discourse: it is written in third-person passive voice, it is presented as factual, and no narrative is used. Instead, focus is on the "disease" or "cancer cells."

As an example of the discourse of alternative cancer treatment, I chose a Web site[8] that advertised and advocated the use of Tian Xian liquid, developed in China by Wang Zhen Guo, for the treatment of lung cancer. I consider the discourse of this Web site to be "alternative" or "complementary" because it is an "intervention not taught widely in U.S. medical schools or generally available in U.S. hospitals" (Eisenberg quoted in Lutz and Fuller 1998). In addition, the site contains—and in fact highlights—elements not characteristic of traditional Western medical discourse, such as testimonials and an award gallery. The Tian Xian liquid itself is described as "an alternative dietary food supplement to help destroy, control and inhibit Cancer by strengthening the body's immune system"; the liquid consists of essences of roots and stems of various herbs and plants such as ginseng (see http://www.tianxian.com/english/ingredients.shtml).

I chose cancer because it is an especially well-known condition, or set of conditions, that receives a great deal of public support in the form of, for example, tax dollars. (Indeed, for fiscal year 2001, NCI has asked for $3.5 billion [http://www.nci.nih.gov/legis/fy2001.html].) Although comparatively few children or adolescents are stricken with cancer, many students know of an older relative or family friend who has battled the disease. In addition to (or perhaps because of) its being widespread, I chose cancer because a large number of alternative and complementary therapies are practiced to treat the disease. A great deal of information on alternative and complementary treatments is readily available; one large-scale study found that nine percent of cancer patients use these therapies instead of or in addition to conventional treatments (Lerner and Kennedy 1992). These therapies have received widespread attention and mixed reviews; although treatments such as laetrile have generally been dismissed by the mainstream medical community, a 1997 National Institutes of Health consensus conference endorsed the use of acupuncture to treat side effects of chemotherapy (see http://cancernet.nci.nih.gov/).

I chose Web sites rather than print sources for two reasons: first, students are, of course, much more likely to use the Web for research than they are to use print sources. Second, and less obvious, I think that it is oftentimes easier to gain information about the organizational context in which the information is presented (and hence answer the questions more completely) from a Web site than from a print source. For example, a student looking at the NCI site on the treatment of non–small cell lung cancer can easily move to other screens that contain information about the National Cancer Institute; however, a print copy of the same information obtained via the NCI's CancerFax service allows no such opportunity.

I developed a set of four questions to ask students about each Web site. These questions grow out of *(a)* composition-based work in cultural studies, such as that by Berlin (1996), and *(b)* work that ties together rhetoric and technical and scientific communication, such as that by Ornatowski (1997) and Halloran (1978). The questions (discussed elsewhere in more detail) are based on the following cultural-studies premises:

- Institutions, operating discursively, impact individual consciousness and action significantly
- Individuals, in their daily lives, attempt to make sense of multiple and competing discourses produced by these institutions.
- Individuals ascribe a level of credibility to the institutions on the basis of the perceived quality of the discourses and on the basis of previous experience with such institutions, if any.

The basic premise that links cultural studies with rhetoric—and, in this case, the rhetoric of medical discourse—is that cultural institutions operate discursively; thus, rhetoric plays an unmistakably critical role in the advancement of these institutions. Ornatowski (1997), in his chapter, "Technical Communication and Rhetoric," explains that the discourses of science and technology are often used for, among other things, the following purposes:

- To coordinate the activities of groups and individuals
- To promote the progression of institutional tasks
- To control activities
- To monitor output
- To make an individual's work advance the established objectives of an institution

I read these purposes as typical of those of any cultural institution's discourse; indeed, they can be understood to correspond to Foucault's (1977a) description in *The Order of Discourse* of the ways discourse is controlled. In this work, Foucault says that control of discourse is achieved by, among other things, the limitation of discourse by disciplines. Disciplinary discourse, says Foucault, must "address itself to a determinate plane of objects" and "be able to be inscribed on a certain theoretical horizon" to be considered "in the true" (1160). Discourse that does not meet these criteria is excluded from the discipline. The correspondence between Foucault's description and Ornatowski's five purposes then plays out like this: the persons in power who control the content and flow of discourse in disciplines (which are, in the case of science and technology, cultural institutions) in turn use this discourse to perform Ornatowski's purposes. All five of these purposes are innately tied to the maintenance and, perhaps, expansion of cultural institutions through discourse, and they have the potential to affect individuals.

The four questions, which I piloted with students from two other courses,[9] are as follows:

1. Which of the sites do you find to be more credible?
2. What specific features of the sites (rhetorical features such as textual and visual content, document design, and use of persuasive appeals) contribute to your conclusion?
3. What other experiences can you draw on from your life experiences that help you reach this conclusion?
4. What would you do to the sites as a Web editor to boost their credibility?

I sent students the questions and the URLs for the two sites in an email message and asked them to respond in kind.

RESPONSES FROM STUDENTS

Responses from students were revealing and centered primarily on issues of credibility. Credibility was, for the students, inextricably intertwined first with what they perceived as the mission of the two Web sites. Students stressed that the NCI Web site provided information, while the Tian Xian Web site tried to sell a product. (Indeed, one of the students characterized the difference as an audience issue, saying that the NCI site was targeted at "information gatherers," while the Tian Xian site was targeted at "somewhat superstitious people.") The perceived marketing mission of the Tian Xian site immediately caused many students to react negatively to the site; several students mentioned that the ".gov" extension associated with the NCI site URL gave it more credibility than the ".com" extension of the Tian Xian site URL. "The Tian Xian Liquid Web site," stated one student, "is provided by Green and Gold International Exports, which indicates that their interest is in solely making a profit." Another said that the Tian Xian site is "a commercialization for the fears of cancer victims, and may be nothing more than a pyramid sales scheme for a product." Similarly, one student claimed that Tian Xian was simply "an impulse buy for Web surfers" and that he was "suspicious" of the product.

Credibility was also tied closely to content. One student stated that she preferred the NCI site because it contained no shopping cart, as did the Tian Xian site, while another stated that she was impressed by the "skillful use of medical language" on the NCI site. The first student went on to say that he was "skeptical of anything [he sees] on a commercial site, because [he knows] that information such as statistics will get skewed in order to better influence customers." Several students stated that the NCI site is more comprehensive because it contained information on cancer in different parts of the body, and they pointed out that contact information, complete with a toll-free telephone number, is provided as well. In addition, although some students found the Tian Xian text easier to understand than the NCI text, the fact that the NCI site contained a dictionary of terms tended to boost its credibility. Visually, some students complained that the Tian Xian site contained too many pictures and that many of these pictures were of poor quality. The proportion of pictures to text was apparently too high for several students and detracted from the ethos of the site; one said that the pictures were distracting.

The content of the two sites contributed directly to their tone. One student focused specifically on this issue, saying that the NCI site "is kind of reserved, like a wise old person, not shouting in your face [as the Tian Xian site does]. . . . The NCI page looks like . . . an encyclopedia page." Several other students maintained that the NCI Web site was much more balanced in tone—presenting both positives and negatives of various treatments—than the Tian Xian Web site, which was described as "cheesy," as looking similar to a "talk show," and not listing "side effects." The NCI site did receive its share of criticism, though, being characterized in tone as "choppy, clinical, and dull" and as "morbid."

With regard to credibility, students also commented frequently on the perceived trustworthiness of the sources of information contained in the Web sites. For example, the information in the NCI Web site came from "reliable sources that go through a rigorous process of drug testing," said one student. Another insisted that the NCI site "had a lot of good information that was supported by medical proof." Although there was a section of the Tian Xian Web site called "Lab Test Results" that listed several Western-style scientific studies, complete with graphs, tables, and references, many students insisted that the products advertised on the Tian Xian Web site needed to undergo "legitimate" scientific testing and peer review in the U.S. and have results published by an American medical journal. The mechanism by which the Tian Xian products act, they said, needs to be identified, clinical trials need to be conducted, and confounding variables need to be identified. Students objected to the narratives ("Testimonials") on the Tian Xian site, which they saw as the main source of evidence used by the company for the efficacy of the Tian Xian liquid. "Instead of providing facts and figures to support the [Tian Xian] drug," one student said, "they gave emotional 'success' stories." Another stated that "Under the testimonials for lung cancer, there are 9 stories of success, a small number to base an effective treatment on." A second student called the Tian Xian liquid a "miracle elixir" with no proof, and a third complained that the ingredients were not listed (in fact they were). "Factual errors" were also noted: one student pointed out that the Tian Xian site contained the phrase "genes entering cells"—a scientific impossibility because genes are present within cells. This same student, however, noted that translation problems (the Tian Xian site is available in six languages) might have contributed to this mistake. Finally, a student dismissed positive reviews of the Tian Xian liquid reprinted on the site, saying that the newspapers involved—*Manila Bulletin* and the *Philippine Daily Inquirer*—sound "similar to the *National Inquirer*," while another commented that awards given to the manufacturers of Tian Xian and listed on the site were from "non-nationally recognized organizations."

In terms of authorship and credibility, the students appeared to take the authority of the NCI Web site (and studies listed therein) authors for granted. They attributed a great deal of credibility to the sponsorship of the U.S. government for the NCI Web site, saying that the government has access to knowledge and financial resources that the Tian Xian manufacturers did not. Indeed, the NCI Web site designers also ensured that users would get this thought in their minds by placing a banner reading "Credible. Current. Comprehensive." across the top of the CancerNet splash page. The Tian Xian Web site did not fare as well. Several students noted that one author (and the only author affiliated with an American institution) on some of the studies listed on the Tian Xian Web site was a professor at Capital University in Washington, D.C.; these students claimed to have never heard of this institution. (Given that a large percentage of students at the institution at which I teach are from the Washington metropolitan area, I would expect that some of them would know it.)

Other authors listed were affiliated with a Taipei, Taiwan–based institution; this fact did not seem to make the Tian Xian site any more credible for the students.

Not all the news was positive for the NCI site, however. It was faulted for being too "texty" and "looking like a hospital." Another student complained that the site was "lackluster" and "monotonous." And, ironically, the same student who lamented the lack of facts and figures in the Tian Xian site previously described faulted the NCI site for being too impersonal, saying that "I don't think they showed any pictures of people or told any personal stories." Another student shared an aesthetic concern, stating that the site should "use more colors besides blue and white." Finally, one student found the NCI site difficult to get around because of its sheer size and scope.

Perhaps the most surprising aspect of students' reactions to the Web sites was the notion that if they were faced with terminal cancer, the students indicated that they wouldn't hesitate to try the Tian Xian liquid if conventional treatments failed. "If I'm desperate," concluded one student in response to question 3, "this might be something that I would want to try." It seems, perhaps, that entrenched cultural codes can be dismissed more easily if one faces an imminent risk of death.

DISCUSSION AND CONCLUSIONS

Students' respective reactions to the NCI and Tian Xian Web sites reflect a thoroughly entrenched, culturally implemented and maintained belief in Westernized scientific and medical practices. In terms of science, the students are—as am I—especially cognizant and supportive of rigorous testing, large sample sizes, peer review and publication, and reproducibility. The vast majority of students accept and embrace the compartmentalized approach of Western medicine, finding that, for example, the NCI site is more credible because types of cancer corresponding to different parts of the body were much easier to find there.

I used students' responses to prompt class discussion in an attempt to illustrate culturally associated assumptions and contradictions that arose in the responses. For example, no student observed that the treatments discussed in the NCI Web site also cost money—much more money than the Tian Xian products—but that these treatments often may be paid for (that is, recognized by) Western insurance companies. I asked students to consider whether making money in the form of federal budget allocations was important to the National Cancer Institute, to ponder whether or not the makers of Tian Xian were interested in helping people, and to think about the similarities and differences between Tian Xian and Western pharmaceutical companies.

Also in the ensuing discussion, students repeatedly reinforced the sense of credibility they gained from the text of the NCI site. Given the extent to which Western society has become a visually based rather than a text-based culture—at least if we are to believe what we are being told—I was struck by how much students praised the voluminous, often fairly technical, and dense NCI text and criticized the perceived large number of visuals in and the colorful design of the Tian Xian site. One student's written response to question 2, which I used to start this part of the discussion, typifies this belief. According to this student, "The site by the National Cancer Institute looks more professional. There is a lot of text, virtually no visuals. . . . The Tian Xian site looks like the center for the next Olympics. The visuals do not go with the seriousness of cancer. It has too many colors and not enough info./text." On the basis of this comment, it appears that we have a ways to go before a primarily visual, colorful design is accorded the same level of respect given to old-fashioned, boring, black and white text when both are associated with scientific and technical content. In addition, our entreaties as technical and scientific communication instructors for brevity and lack of jargon as characteristics of effective, respected discourse do not seem to be having their desired effect as yet.

Although the discussions with students were intriguing and illuminating, I cannot, of course, claim that students can dismiss thoroughly entrenched cultural codes associated with powerful cultural institutions simply as a result of comparing two medical Web sites and then discussing assumptions underlying their conclusions about these sites; nor should students be expected to subsequently dismiss their beliefs even if they are recognized as being culturally constructed. However, by approaching medicine (as well as science and technology more broadly) as a discursively operating cultural institution in a technical and scientific communication course, students can begin to gain the necessary critical tools to demystify and make sense of the myriad claims and counterclaims inherently a part of these discourses and to begin to understand how the discourses affect them individually. Because health is such a vital part of an individual's subjectivity, it is especially important that students understand the contexts in which the discourses associated with the cultural institution of medicine are produced, especially, for example, in terms of who stands to gain what with respect to political or economic power, of recognizing that results of medical research can often be interpreted in different ways, and of realizing that for every voice heard, there are many other voices that are not.

Additionally, students will write, edit, and design more effectively if they know something about the cultural baggage that discourse—no matter what cultural institution it is a part of—always and already carries. In this particular study, attention to medicine as a cultural artifact does not detract from my goal of teaching students to produce effective, rhetorically successful medical discourse. Indeed, a better understanding of the rhetorical situations of which these discursive events are a part most likely increases the chances for students composing discourse that achieves its desired results. This realization also helps students to predict the ethical implications of their writing, an area that Hartung (1998) has pointed out as seriously lacking in our pedagogy. As a result, they can understand more clearly the mechanisms by which they are influenced and, in turn, gain a more complex understanding of and critical appreciation for the medical discourse.

In this study, students were able to recognize assumptions about Western science and its manifestation as a cultural institution by contrasting an example of Western scientific discourse with its counterpart from a different cultural context. Such a contrast invites a relatively rapid and invigorating response. An intriguing and challenging follow-up to such an assignment would be to ask students to then focus on a piece of Western scientific discourse on its own, without the benefit of contrast to help sharpen the assumptions, to determine if the critical skills they gained via the National Cancer Institute/Tian Xian comparison can still be applied. For example, the instructor could ask students to find some relevant Western research that relates to a health issue that they, a family member, or friend is facing and to develop a series of critical questions not explicitly addressed or answered in the research.

12

COLLABORATING WITH STUDENTS

Technology Autobiographies in the Classroom

Dickie Selfe

A STORY

As this volume suggests, all sorts of instructional innovations are possible. As a result, it is sometimes hard to tell which way to move. Where should we put our energies, limited as they are? In this chapter, I suggest that we might go about productively innovating our technical and professional communication pedagogy and curricula by paying attention to students' literacy skills as they come into our classes and programs. In particular, we might want to attend to their technological literacy skills, attitudes, and approaches to learning. A short anecdote might be instructive.

My observations of student's technological literacies and their impact on technical communication (TC) courses are reflected nicely by the experiences of one of my former colleagues at Michigan Technological University, Dr. Dale Sullivan. Sullivan was teaching a Web design class in the spring of 2000 when he commented on an unexpected turn the class had taken:

> I discovered that the coding ability of the students tended to be very advanced. The problem is that we don't know where students are, and one class in Web design will attract a very wide array of abilities. For 80 [percent] of the class, I should have been teaching perl and cgi script applications and xml and asp. For 20 percent I should have been teaching very basic html and how to put a Web page on a server. . . . So I simply resorted to teaching basic principles of user-centered design, architecture, navigation, user interface design and user testing, and turned the class into a group work operation. (Email correspondence, May 5, 2000)

This chapter suggests that the flexibility and nimble curricular redesign that Sullivan was able to manage in this class is becoming more and more the state of affairs in many technology-rich TC courses.[1] His note caught my attention because the experience so closely matched my observations: there have been surprising fluctuations in technological skills, approaches, and attitudes in student population over the past several years.

PRACTICING WHAT WE PREACH

It's a good day when students write back to their home institution and let teachers know how their lives are going. Several years ago, one such message came across my screen, and I've used it ever since in all of my TC courses. It came from a young woman who had graduated a few years before and for whom I had great respect. I asked her about the most important three lessons she had learned in her tenure as a technical communicator. She said, "Know your audience; know more about your audience; and really get to know your audience." I've used her quote many times in the classes I teach.

We ask students to know their audience at every turn in the technical communication curriculum, occasionally providing them with methods for such analyses. In light of the changing technological experiences that Dr. Sullivan encountered in his course and the obvious need for communication specialists to know their audiences, I began to realize that I should have been following my advice. After all, as a teacher, I am continually constructing learning "interfaces" for students: interfaces that consist of online environments, content material, composing processes, in-class activities, and so forth. It occurred to me that I should also be making a stronger effort to know these students and to construct learning experiences *with* them, not just *for* them.

This chapter proposes a method—an autobiographical assignment focused mostly on past technological experiences—that might well benefit students and our class and curricular planning. As assignments, technological autobiographies (TA) are wonderfully functional. They provide an interesting glimpse into the attitudes, experiences, learning strategies and levels of expertise that students bring with them into our classes. They are writing samples; they are introductory narratives that help form our understandings of each other as people, workers, and learners. They help us and students get to know, know more, and "really" know an audience that we often use in our TC classes: the class members themselves.

But, in addition, these assignments can be part of a participatory design "method": technology autobiographies are windows into student lives in an age of rapidly changing technologies, technologies that have become central to our educational and professional endeavors on and off campus. I am certainly not the first to claim that we need to adopt and adapt user-centered, participatory design methods to the design of classes (Soloway 1994). In his article about designing online courses, Stuart Blythe (2001) makes it clear how difficult the academic setting makes this method. His solution to these difficulties comes in two forms. First, we have to imagine, as usability advocates have for years, that end-user (student) participation in the design of our classes is ongoing and formative. It should be an expected component of the ongoing redesign of our classes, not a one-time usability event. Second, he provided opportunities for each generation of students taking his class to choose the focus of projects in its class. Thus, his "assignment" becomes part of a participatory design method that will inform not only the class at hand but also the next improved, technology-rich, instructional experience that he helps construct.

THE TECHNOLOGIES IN OUR DISCIPLINE

But students' changing technological literacy practices aren't the only reason for adopting participatory design methods. As I've suggested in other publications, our culture is currently experiencing an overdetermined state of technological change. This change is particularly true in our discipline (see, for example, R. Selfe 1998). One graphic method of characterizing the overdetermined nature of this state of affairs is through the guillotine chart originally constructed by IBM in 1979, then revised by Dwight Stevenson in 1984 and Roger Grice in 1987. No doubt the blade has gotten longer and sharper since then.

Roger Grice, in his 1987 dissertation *Technical Communication in the Computer Industry: An Information-Development Process to Track, Measure, and Ensure Quality,* describes how in 1979 the IBM Human Factors Task Force met in Atlanta, Georgia, to "discuss and chart future actions" (50). Among other actions, they defined the role of technical communicator as "information developer" and created a guillotine chart much like the one presented in figure 1 that shows the growing responsibilities included in that role. In 1979 only the first four columns were represented. In 1984 Dwight Stevenson's prescient expansion of the chart showed the

Information Development Job Description and Direction

1960 → 1985

	Designed After	*Designed With*	*Designed as an Integral Part*	*Systems Design*
Purpose	Product Description	Functional Description	Task-Oriented Use Description	Process Design and Description
Development Emphasis	Content	Development and Schedule	Field and Customer Cost	Product and Process Evolution
Objective	Completeness	Technical Accuracy	Ease of Use and Total Cost	Efficiency, System Optimization
Product	Books	Libraries	Information	Information, Especially Electronic
Volume	Low	Medium	High	High in Volume, Complex in Nature
Skills	Document Writing / Editing	Computer Hard/Software Engineering Writing Editing Planning Graphics Text Management Testing for Accuracy	Retrievability Writing Editing Planning Graphics Management Text Management Testing for Accuracy and Usability Audience Definition Measurement Financial Human Factors Media Packaging Distribution Publishing Task Definition	Substantive Editing Substantive Writing Graphics Layout and Production Video and Film Training Interpersonal Communication Organizational Behavior Planning Text Management Computer Text Production Testing Financial Analysis / Management Product Distribution Product Development Online Documentation Computer Graphics Information Research Database Design and Management Legal Protection Cross-Cultural Communication Software Development Software Management Public Policy Research Design Research Synthesis

Figure 1. Grice Guillotine Chart (adapted from Stevenson 1984)

technical communicator "moving into the area of system design, especially design of the user- system interface" (Grice 54). It's hard to argue that Stevenson wasn't entirely correct in his assumptions that information developers would soon be engaged in video and film production, product development, database design, and software development, to name just a few of the additional "skills" suggested by his chart. And this, of course, was before the meteoric rise of the World Wide Web as an interactive, multimedia information delivery device.

PARTICIPATORY DESIGN AND THE TC CURRICULUM

If anecdotal evidence (Sullivan and Blythe, for example) suggests that changing technological literacies will or should change the courses we teach and if the increasing technological complexity of the discipline itself will encourage us to adapt our courses, one might then ask, "What do we mean by participatory design?" I would suggest that it is more of an attitudinal change than any one particular method. That attitude will then lead us to innovative pedagogical approaches and implementations. In a special issue of *Human-Computer Interaction* on "Current Perspectives on Participatory Design" (PD), Randall Trigg and Susan Anderson (1996) suggest a common theme among the many approaches found in this design rubric. In PD, there is "a fundamental respect for the people who use technology and for the right of people to have a direct influence on decisions that affect their lives" (181). Changing technological literacies are so fundamental to the TC curriculum that we will probably find traditional usability methods—focus groups, questionnaires, controlled usability testing—useful but "not sufficient to the development of genuinely *useful* systems [in this case, educational systems]" (Blomberg, Suchman, and Trigg 1996, 239). As we design technical communication programs or classes, we are in essence aiming at a moving target, one moving on several dimensions at once. At least two of those dimensions seem obvious: students bring a rapidly changing set of technological literacies practices into our classrooms each term, even as the technologies we are asked to use change around us.

As a result, we need to rethink our relationship to students. They are, after all, the workers in an educational system, a system that is, in my experience with the program at Michigan Technological University over the past fifteen years, constantly in the process of being redesigned. TC academics face a growing list of skills designated by commercial representatives and changing theories of the communication process. One might reasonably ask, then, another question: "Isn't it enough that TC teachers and curricula designers consider the suggestions of theorists, industrial advisory boards, employers, and technology experts in the redesign of our programs and classes?" The answer for those working on technical literacy projects is, of course, "No."

In a research report submitted to the Society for Technical Communication in March of 2000, called "Studying the Acquisition and Development of Technological Literacy," Cynthia Selfe and Gail Hawisher summarized the problem we face this way:

> So, here is a problem: We know very little about how and why particular individuals acquire and develop, or fail to acquire and develop, technological literacy.
>
> And, here is another problem: We know very little about how large-scale historic, cultural, economic, political, or ideological movements act and interact to shape individuals' acquisition and development of technological literacy[3]—or how individuals' literacy practices and values help constitute these macro-level trends.
>
> And here is a really big problem: Despite our lack of knowledge about these important matters, literacy experts, educators, and policy makers continue to set standards for technological literacy (*National Educational Technology Standards for Students,* 2000; *Standards for the English Language Arts,* 1996), create educational and workplace policies about technological literacy (*Getting America's Students Ready for the Twenty-first Century,* 1996); and design programs and curricula that teach technological literacy in schools and in the workplace (*National Educational Technology Standards for Students,* 2000). In sum, we're basing big decisions on minimal information. (1)

As suggested in Selfe and Hawisher's comment, TC curriculum designers, myself included, have habitually relied on system-centered approaches as they face the escalating curricular requirements driven by diverse skill sets like those represented in Grice's guillotine chart (see figure 1). We attend to our favorite theorists, available industrial representatives, and technology specialists, but rarely the student populations who inhabit our programs. We "black box" their changing literacy practices at the risk of becoming increasingly irrelevant or at least disconnected not only from students and their learning habits but also from the youth culture in general.

THE AUTOBIOGRAPHIC ASSIGNMENTS

The story that Dr. Sullivan tells about his Web design class strikes a chord with me because my ability to predict what students will bring into the class has likewise been unsettled. I have clearly overestimated and underestimated their abilities in the past. At the same time, literacy scholars like Deborah Brandt have come to some interesting conclusions about changing literacy expectations. In "Accumulating Literacy: Writing and Learning to Write in the Twentieth Century," Brant (1995) interviewed sixty-five participants with the goal of discovering the "institutions, materials, and people" that inform the acquisition of "practices that haunt the sites of literacy learning" (651, 661). One of her findings suggests that there is an increasing "escalation in educational expectations" on literacy practices both in the home and in the workplace (650) as a result of recent, incessant technological "innovations." The technology autobiography and related assignments developed in this chapter are generally beholden to scholars like Brant. More specifically, they have been developed in detail and practice by my colleagues Karla Kitalong and Michael Moore. In a forthcoming publication, we speculate that "these heightened expectations [are] articulated by a wide variety of educational stakeholders, including the media, state legislatures, industry, and any number of special-interest groups." Our approach highlights "the contradictions and ambiguities between institutional goals and the communicative acts and literacy practices of students, articulated in their own words" (Kitalong, Selfe, and Moore forthcoming).

In the following section, I discuss how I instituted versions of the autobiographical assignment in classes with two different populations of technical communication majors: one set of assignment responses came from junior and senior undergraduates and the other from masters-level graduate students in a different professional communication program. In both cases, the course name was the same: Publications Management. As you might expect, each set of responses to the assignment taught both me and the students a slightly different lesson. For that reason, in the next section, after describing the assignments themselves, I'll explore possible implications gleaned from each collection of responses. The implications I drew from these student reflections provided the impetus for immediate classroom innovations and were valuable as I planned future courses and programmatic proposals for modifying the undergraduate and graduate programs in which I work.

The process I used in both classes (graduate and undergraduate) was similar:

1. I assigned the autobiographical activities described later.
2. I combined their reflections into a class booklet (hardcopy and .pdf versions) to be used as a text in the class.
3. I asked students, after reading the class booklet, to speculate on the range of learning styles, attitudes, technological skills, and experiences (LATEs) they saw in the combined document. (I also participated in this process.)
4. I worked with them to determine how these LATEs should influence the technology modules (instructional documentation) they would be producing for the class.

For the purposes of this chapter, I'll touch briefly on step 4 but focus primarily on the value of step 3. I, of course, learned at least as much as the students from their responses to this assignment. Students were informed that the autobiographies were also part of a classroom research project that would be used to help reconfigure this and other TC courses and used to make recommendations that I hoped would influence technical and professional, undergraduate and graduate communication programs in the future. The technology autobiographies, then, had two primary purposes:

1. To help the students learn more about an audience that they would be addressing in future assignments
2. To help me better understand students and their relationship to technology and course content

In both classes students were asked to respond to the following question sets in informal, autobiography assignments.

Questions Leading to the Undergraduates' Technology Autobiographies

1. Write and/or draw an autobiography in which you recall your earliest experiences with technological devices or artifacts. What were they? What do you remember about using them?
2. What were the popular gadgets in your house while growing up?
3. Who[m] do you identify as being most technologically "literate" in your life?
4. What's on your desk at home?
5. What technological devices are you carrying now?
6. What's on your technological "wish list"?
7. How do you expect to deal with new technologies in the future?

8. What sort of documentation works best for you?

(Artistic representations [are welcome] and need to be accompanied by a written statement explaining the work.)

Notice that the eighth question—"What sort of documentation works best for you?"—was not directed so much at the technological portion of this assignment as it was at the content of the course. The upcoming assignment would ask students to design, test, and produce a technology documentation for the people using a local computer lab. You will see that I expanded this section in the next iteration of the autobiography assignment (for the graduate class). I included an entirely new focus for student reflection: the publishing autobiography. I quickly realized that it would be useful to know more about students' past experiences not only with technology but with the course content as well. In both cases, I hoped to find ways of involving students more intimately in the design of assignments and products that come out of the class. A version of the assignment as I explained it to students follows:

Technology and Publishing Autobiographies

These two pieces of writing are meant to be fun and interesting to your classmates and myself. We need to know a bit about you. I would like you to write a personal technological autobiography (TA) and a publishing autobiography (PA). We'll start this assignment in class and then, after you complete your autobiograph[ies], distribute them electronically (in .pdf format) as a booklet. The class will have the weekend to read them. It's a lot of reading but by the second class, we will all know something about each other and our collective technology and publishing experiences. These TAPAs will NOT be graded other than to note that you handed them in. . . . To complete the assignment, respond to the following prompts:

Technology Autobiography (TA):

Write and/or draw an autobiography in which you recall your earliest experiences with technological devices or artifacts. What were they?

- What do you remember about using them?
- What were the popular gadgets in your house while growing up?
- Whom do you identify as being most technologically "literate" in your life?
- What's on your technological "wish list"?
- How do you expect to learn and keep up with new technologies in the future?
- What technological workshops are you willing to develop for me and your classmates?

- What technologies do you need to learn in the near future?
- (Artistic representations of your relationship to technology are very welcome and usually very interesting. I would appreciate a short written statement explaining the work.)

Publishing Autobiography (PA):

- What experiences in your past have gotten you excited about publishing?
- What informal or formal (work-related) publishing experiences have you had?
- What specific publishing expertise do you bring to the class: organizational, audience analysis, technological, experience with types of publications, . . . ?
- What sort of publishing NOW interests you? In other words, imagine yourself working for a company, organization, or start-up that you really believe in: describe what kind of work they would do and what sorts of publishing they would engage in.

Write both the TA and PA with your classmates' interests in mind. What examples would be most interesting to them? How much time do they have to read about your experiences? What do they need to know about you and your abilities?

Why start with this information?

Almost every professional/technical communicator I talk to about her/his job mentions the need to know more, more, and still more about the audience being addressed when creating a publication. In other words, knowing your audience very intimately is more important to a successful publication than almost anything else. Your first individual project will be to construct a technology module for users of your home computer lab. And if my experience holds true, even those of you who have worked in this lab know very little about the literacy skills and learning habits of those around you.

IMPLICATIONS FROM AUTOBIOGRAPHY ASSIGNMENTS IN UPPER-DIVISION AND GRADUATE TC COURSES

The Technological Ambivalence in an Undergraduate TC Course

In both classes, I was introducing students to the publishing industry and, in that process, relied on real client projects. Because of the need to use imaging and publishing software and hardware in both classes, it seemed important to identify what students already knew and what they could add to the class (because many students came in, as was the case in Dale Sullivan's class, with skills more advanced than the teacher's). I also wanted to better understand what attitudes and learning styles they had adopted in the past. The technology autobiography assignment for the undergraduate class was one way of collecting this type of information and incorporating it into the planning of subsequent courses and sessions within this particular course.

Learning About Each Other: An Aside

Because of a subsequent assignment, it was important that these students learn a great deal about each other, and the autobiography booklet provided that opportunity. The assignment asked them to develop "technology modules" (instructional documentation) that would be useful to students working in the drop-in lab that they frequented. The students in the class would be, as a result, both the creators of helpful technology modules and representative users of those same modules. Not surprisingly, these informal autobiographies were remarkably useful in our audience analyses ("really get to know your audience") for this assignment. After receiving their short autobiographies, I constructed a single document that contained the entire class set (thirty pages long). That booklet became a reading assignment out of which the students were asked to develop a user profile for their technology modules. The technology autobiographical document gave us the exigence for discussing the nuances of audience needs, expectations, and learning styles at a level well beyond the generalities I often received in students' previous audience analyses.

But the autobiography assignment gave me and the students information about the technological literacy makeup of the class that seemed just as or more valuable at pedagogical and curricular levels.

Perhaps because these students were young, burgeoning professionals and just beginning to realize the full extent of what communication technologies would mean to them in the future, this set of autobiographies, as a whole, illustrated the ambivalence that students have for their technology-rich futures. In a future publication, Karla Kitalong, Michael Moore, and I discuss this ambivalence in more detail (forthcoming). Here, I'll summarize some observations that seem to have implications for TC classes and programs in the midst of pedagogical and programmatic change.

A Diversity of Experience

One observation common to all the sets of technology autobiographies that I have reviewed over the past two years (our research team and others have been asked to apply this type of assignment to a number of technology-rich English-studies classes) is that students bring a wide range of technological experiences to bear in the TC classroom. This might be best illustrated by a student's description of what I call generational compression of technological experience:

> My little brother, who is four years younger than I am, just graduated from high school with more knowledge of computers and technology than I will ever learn. He just built himself a computer from scratch and is currently attending [XXX] State to study computer networks and systems. My sister, on the other hand, is only two years older than I am. She spent 4 years in college without ever having to turn a computer on. (Paula, pseudonym)

Students realize that there are radical differences in experience levels. Those variations in experience, however, don't necessarily reduce the opportunities for hard-working, self-motivated students, as the next quote indicates:

> [A] calculator was my only real link to technology [in high school] until I managed to actually touch computers again in college. I was overwhelmed when I got here. I had no clue what computers could do. At the time I was an electrical engineering major. Now, I'm a computer science major [and one of the most technologically adept students in the class]. (Otto)

But for the average student entering our technology-rich programs, we can't assume that they will all simply "catch up" magically. They worry about, and we should worry about, how our classes might better facilitate the catch-up process, and, at the same time, we should continue to challenge those students who come in with Paula's brother's level of experience.

A Backgound of Gaming

A second component to the technological ambivalence in these undergraduates' TAs is a growing experience with gaming systems. In all sets of technology autobiographies, educational and purely entertainment-level gaming has a strong representation. To a follow-up question to students who claimed a strong gaming background came these responses. "Believe it or not, games can make children less frightened of technology. I thought of computers as a toy for years before it actually became a tool." (Johnson, email correspondence, Sept. 22, 1999) Not only did they suggest that gaming reduces computer anxiety, they hinted that specific learning strategies were encouraged by games. The following is an extended quote from a young woman's reflections on gaming:

> What gaming taught me is that there are always little tricks to doing things. For example, when i played supermario bros. i learned how to "warp" to different worlds and that meant that i could skip 4 levels of playing without losing points. So i would always try to do new things regardless if there was a hint that i could do it or not. . . . The hidden shortcuts really got to be fascinating. . . . But what is also key is that i learned a lot of tricks from my friends. . . . So that is getting a reward [from others'] experience with the game. (Glenda, pseudonym, email correspondence, Sept. 22, 1999)

These two short comments suggest that students with gaming experiences might be more willing to approach new technologies fearlessly, try techniques regardless of whether those techniques seem possible or not, and seek out shortcuts and tricks on their own and with friends. The questions that come to mind first include the following: How can students, who are designing instructional modules, or how can we, who are devising technology-rich classroom activities, take advantage of this playful, exploratory attitude? Will someone with this type of background approach technology instructions or our classroom activities in interesting and unique ways? As this type of gaming experience becomes more common than exceptional (nine of nineteen students in 1999 claimed to have had substantial gaming experiences), how will our approach to online and print-based learning systems and documentation change? How should our approaches to teaching change? What will this mean for technical and professional communication departments at a programmatic level? The ambivalence here resides mostly in my concerns, not students'.

Learning Styles Differ Radically

The technology autobiography assignment asks explicitly about how students learn or plan to learn new technologies; this led to the third component: students' technological ambivalence, which came out of the undergraduates' reflections and which has to do with radically differing learning strategies. Though a certain percentage of students have a tinker's mind-set—one that encourages them to understand the underlying workings of the technologies they use—most admit to short-term, just-in-time learning patterns that allow them "to stay current with those things that pertain to my field or are positioned in it" (Randall).

Students will apparently come to us, not surprisingly, with a number of learning strategies, some of which won't be a comfortable fit for many of technical and professional communication teachers: students will be crisis learners; fearful, reluctant learners; stealth learners willing to make the trade-off between their depth of understanding and the practical art of getting the job done. They are also aware of the trade-offs they may have to make if they are going to commit to learning new technologies thoroughly.

> I have a hard time throwing off other, maybe older, values for the sake of my computer literacy. I recognize that it takes a tremendous time commitment to stay fluent. I don't know what other part of my life I want to give up so that I can learn yet another piece of software. I will probably manage the learning of future skills by crisis, doing only what I have to do to remain literate enough. (Diana, pseudonym)

These comments only hint at the ambivalence students sometimes express about their future with constantly changing computing technologies. They sometimes speak about enslavement, painful values, reluctant learning, impersonal lifestyles, and rude online behavior. All are words and phrases that make it clear why students might approach our use of communication technologies with some reluctance. Our job, as a result, would seem to go well beyond the introduction to new—sometimes useful, sometimes painful—bleeding-edge technologies. Technical and professional communication instructors may well need to begin asking themselves what strategies they themselves adopt to stay reasonably and appropriately current. More difficult might be to imagine innovative approaches that make those strategies an explicit part of our instruction.

To summarize, this one set of technology autobiographies led me to reimagine several components of TC courses and programs. The first is to question how we accommodate the wide (some would say "ever widening," C. Selfe, 1999) technological experiences that students bring into our classes. The second is to imagine how the substantial online experiences with gaming systems will change the way students work and learn in our classes. The third is to build a robust set of strategies for adopting new technological systems into our curriculum and for adapting to them. These concerns are not necessarily going to emerge in all technical and professional communication courses. The autobiographies do, however, give me data and an agenda to take back to our curricular committee, which is endlessly reconfiguring and reconstructing the requirements and courses in our program. I assume that other programs, having collected their data, might well come up with unique (innovative) concerns that they might address in their curriculum as well.

One of the most interesting aspects of collecting these autobiographies over time and courses, however, is that patterns begin to appear. As I read through the technology and publishing autobiographies (TAPAs) from a graduate professional communication course, not only were patterns evident, but potential pedagogical solutions also seemed to present themselves in the anecdotes and descriptions students provided.

LEARNING FROM THE TECHNOLOGY AND PUBLISHING AUTOBIOGRAPHIES (TAPAS) FROM A GRADUATE PROFESSIONAL COMMUNICATION COURSE: REASSESSING THE COURSE ITSELF

Content Feedback: Another Aside

You might have noticed that the second autobiography assignment added questions aimed at better understanding not only the technological literacy of the class but also the content experiences of students (in this case, publishing autobiographies that detailed their experiences with print publications). In future classes, I plan to use the content questions to help organize that portion of the class as well. Students' varied publishing experiences in this set of autobiographies have convinced me, for instance, of the value of several strategies:

1. I might ask students to recruit expert consultants to visit the class physically or virtually during the term.
2. Productive interviews with working professionals (again either face-to-face or virtually) will enrich the class and subsequent classes.
3. It should be possible to develop sustained relationships with some of these professionals to set up lively online discussions that provide professional relevance and contact with industry professionals even in the remote north woods of Michigan Technological University.

But the more generalizable and important innovations for technical and professional communication programs would seem to be located in the technology autobiographies of these students.

Adopting and Adapting to New Technologies

Although there is always a great deal to glean from the technological experiences that students describe in their autobiographies, each set seems to provide some insight into particular or related portions of the courses I teach. The informal TAs assigned to this graduate publication course in the fall of 2000 made it clear that students from both courses were quite concerned about one issue in particular: the need to develop strategies for adopting and adapting to new technological systems.

> I read these journals and magazines advertising new software and technologies and wonder how on earth I am supposed to stay competitive in the job market when I haven't learned the old stuff before they introduce the new. (Brown)

Ms. Merrill says,

> I did eventually get my hands on a word processor, which was replaced by an actual computer when I was starting college.
>
> What do I remember about using them? Fear. Trepidation. I remember being more than a little daunted by those computerized thingys with all their buttons and options and programs. This reluctance did not last long, however. Without ever really consulting the instruction manuals except in extreme desperation, I figured out how to work them through guess-work and trial-and-error.

Statements of anxiety seem common to many students coming to technology late in their academic careers. The lucky ones are thrown into an institutional setting where they are required to learn new systems quickly to survive. Usually, they have no structured way to develop any systematic method of learning new technologies and so are thrown back on survival strategies: just-in-time learning, trial-and-error efforts, guesswork learning patterns. These are the lucky ones. Unlucky students may not even have these opportunities. Students' anxiety was compelling possibly because theirs reflected mine so accurately. They made me wonder whether together we couldn't devise some alternative models that would better serve their technological literacy needs. As I read their stories, I began to imagine some possibilities.

Resources and Procedures for Staying Abreast

> I try to stay educated about technological trends, though. I peruse magazines, engage in conversations about what's "out there", and play enough video games to understand the implications of new technology to the entertainment industry. To this day, I still relate my experiences of technology through my obsession with entertainment, because I have learned the most about technology through this medium, and I honestly think entertainment drives technology more than any other single factor. (Bonhan)

Obviously, Bonhan hasn't been exposed to high-end military applications or he might change his mind slightly. But his claim about gaming might also be closer to the truth than I might think. His comment suggests to me that we might incorporate explorations like the ones he mentioned during class by asking students to find reliable sources for technological news relevant to the class and share summaries of their visits (to Web sites), articles, posts, and so forth with the class. If one of Bonhan's techniques is to "play enough video games to understand the implications of new technology," then we might ask him and other gaming aficionados to bring the implications they draw from that experience to class. If TC instructors make this type of activity an explicit part of our exploration, we might all be able to imagine how the gaming industry can provide some positive, productive protocols to technical communication professionals interested in instructional systems and online interaction.

Setting Up Collective Learning Experiences

One approach to reducing anxieties and providing students with strategies for learning new technologies is to set up appropriate technological learning experiences within our classes. The sixth question for the masters-level class was supposed to help me set up this type of learning experience. (What technological workshops are you willing to develop for me and your classmates?) Surprisingly, even those experienced with some technologies often responded this way: "At this time, I don't feel that I have enough expertise to lead a technical workshop" (Julie).

Students are justifiably concerned about teaching their peers. If public speaking is one of the most anxiety-producing events in a person's life, consider how nerve-wracking public teaching must be! Students often don't believe that they are *the* most technologically savvy person in the class and so feel incapable of leading a technology session. I haven't given up on this approach, however. Instead, it has become my job to convince them of the *kind* of workshop they can develop and lead, not *whether* they can lead one. To pave the way, I might

1. provide them with a model interactive learning situation appropriate for the class,
2. survey them and identify the expertise in the class around which they can form teams,
3. help them construct interactive learning modules that will help bring the entire class (including the teacher!) up to some baseline understanding of a particular software, hardware, or netware environment; we could then . . .
4. implement those learning activities systematically during the course.

Not only will we all learn a great deal about relevant technologies, but these activities are also full of opportunities for typical technical communication compositions (textual, visual, aural, animated, and so forth), oral presentations, interactive instructional presentations, and the development of online help systems and Web-based constructions.

Technological Mentoring Programs

> The second event that changed the way I look at technology occurred when I became friends with the guy I am now dating. He was raised in a technologically advanced environment, so computers are quite the norm for him. (Loftin)

Though we are unlikely to intentionally set up close personal relationships between students, there is every reason to believe that it might be productive to add a mentoring experience to capstone classes for graduating seniors. A technological mentoring assignment may provide a way to harness the expertise of seniors and pass it along, at some level, to our younger students. On the other hand, if the technological compression described earlier holds true, these graduating seniors might also find themselves learning a great deal about new technologies from the younger students. And they will certainly learn a bit about relating to younger colleagues, a skill of some significance in the workplace.

For years, I've encouraged teachers in professional development workshops to recruit their best, most enthusiastic former students as technical resource collaborators. These students can help implement a version of the technological mentoring program within a single class. Though it takes some preplanning and organizational finesse, it is possible to recruit former students to work with your current class as volunteers or for small stipends or for independent study credit. I've seen even the most technologically savvy teachers use mentoring programs within their classes successfully.

Stand-Beside Consultants

An idea closely related to the technological mentor is the stand-beside consultant. When asked what software or hardware they might explore with the class, surprisingly, most students expressed unease at the prospect of leading a technology demonstration for their peers. But they were quite willing to offer help to novices as they worked. Programmatically, we might want to consider in-class or out-of-class activities that pair those new to specific systems or software with experienced students as they work on class projects. Using these students as "stand-beside" consultants will give consultants valuable teaching experience that they will no doubt need in the workplace. Both novice and consulting participants can be asked to produce reflective writing that details the kind of collaboration and consulting that worked well. Collected, these reflections could be combined into a "text" to be read and discussed by the whole class.

This begs the question, however, of how to push the more experienced students in the class. Perhaps we should be making time in the course for technologically advanced students to push their skills along by collaborating with out-of-class consultants if the teacher can't provide the expertise. We might contact former students, local professionals, or students from other majors as we attempt to recruit these consultants. Part of the assignment for advanced students could be to create a learning experience for the rest of the class that will introduce us all to the new systems they will be learning.

CONCLUDING STATEMENTS

The Method

A great deal of the "method" involved in applying participatory design processes to our classes and programs has yet to be developed. I look forward to dialogues with other TC professionals as we try to imagine how learner-centered design might best be applied to the technology-rich instructional environments.

The Work Load

Resource assignments, collective learning experiences, mentoring, and stand-beside consultant programs are all the more work for the teacher or more planning for the program director. As such they have to be weighed against the already substantial responsibilities of teachers, administrators, and students. As this volume makes abundantly clear, all innovative pedagogies seem to carry the same onus: they are typically more work than the status quo. But if we have to make choices about how we manage our courses and curricula, technology autobiographies at least give us vivid and contextualized "data" to use as we challenge our traditional pedagogical practices in a changing technological landscape.

The Context

As I mentioned early, the speculation about how we might draw on the observations from technology autobiographies to redesign classes and curricula is still quite preliminary and must be placed in unique programmatic contexts. What seems appropriate in a TC program at a technological university located in a rural, isolated region of the north woods (Michigan Technological University) and what seems essential to students living near the research triangle of North Carolina (Clemson University) are quite different. I look forward, however, to exploring the patterns of common student experiences with colleagues from around the country (and worldwide) as a prelude to innovating and making relevant technical communication programs of the twenty-first century.

PART THREE

Pedagogical Partnerships

13

A PEDAGOGICAL FRAMEWORK FOR FACULTY-STUDENT RESEARCH AND PUBLIC SERVICE IN TECHNICAL COMMUNICATION

Brad Mehlenbacher
R. Stanley Dicks

It has become a truism that students learn substantially more by augmenting their traditional education with collaboration and hands-on activities, particularly with activities that they feel ownership for or that they perceive to have real-world relevance (Honebein, Duffy, and Fishman 1993; Rogoff 1990; Savery1998). Of the "seven principles of good practice in undergraduate education" listed by Chickering and colleagues (1987, 1998), for example, the majority involve motivational and "social" and task-based dimensions for learning. Thus, effective instruction encourages student-faculty contact, cooperation among students, and active learning; gives prompt feedback; emphasizes time on task; communicates high expectations; and respects diverse talents and ways of learning.

It therefore seems reasonable to argue that students would benefit from working with faculty in cutting-edge research because it may provide them with their first opportunities to make real contributions to the professional literature in coauthored publications with faculty. Moreover, getting students directly involved in real research gives them a richer understanding of their chosen discipline and involves them early on in their careers as potential contributors to the field.

In this chapter, we describe a unique collaboration among university administrators, faculty, students, and constituents that originated in a proposed project to conduct usability testing for a part of the university's Web site. From October 1998 to June 1999, the authors were charged by the North Carolina State Extension, Research, and Outreach Office to usability test *Ask NC State,* the extension branch of the NC State Web site and a creative presentation of the numerous online resources available for potential extension audiences. The problem driving our initial conception of the project proposal was how to integrate our teaching, research, and service goals into a multidimensional, collaborative effort that would produce benefits in each domain and for the various audiences (or shareholders) involved. The solution was to have students across three classes, acting as research apprentices, help us perform usability tests with various constituencies of the university who might have reason to access the *Ask NC State* Web site.

In this chapter, we describe how we achieved the following results:

1. Provide students with access to cross-disciplinary perspectives in cognitive psychology, human factors, computer science, industrial engineering, and technical communication
2. Provide students with hands-on experience performing usability testing on a significant Web site
3. Involve the university's Extension, Research, and Outreach program in the pedagogical goals of graduate and undergraduate instruction
4. Perform research related to the usability of Web-based materials
5. Contribute to improvement of a part of the university's Web site
6. Involve students in analyzing, compiling, and presenting research information for an authentic audience

Our ultimate hope was that the project would serve as a blueprint for future research projects for several reasons. The grant

allowed faculty-investigators to integrate their extension, teaching, and research goals into a single project;

supported both undergraduate and graduate students' efforts to work closely with faculty in the evaluation and improvement of an official NC State function;

encouraged the involvement of potential extension audiences, serving as test participants, in the development of materials designed to support their information needs related to NC State;

gave credibility to the importance of usability testing and evaluation as an integral part of the creation of materials for use by extension audiences; and

served as a possible model for future collaborations among Extension, Research, and Outreach, faculty, students, and, ultimately, industry partners.

Because the project began prior to the semester, collaborating faculty were able to design and implement usability plans for working with the students to test the targeted Web site and to carefully design their syllabi to integrate the extension project.

PEDAGOGICAL/THEORETICAL FRAMEWORK

Tebeaux (1989) and Zimmerman and Long (1993) have argued persuasively that teaching students how to collect, analyze, and report data to various audiences and with different purposes should be a chief pedagogical goal for technical communication instructors.

Though many technical communication curricula have certainly adapted this perspective, many educational theorists still advocate dramatic educational reform. Koschmann, Kelson, Feltovich, and Barrows (1996), for example, cite dozens of studies, revealing that "existing educational systems are producing individuals who fail to develop a valid, robust knowledge base; who have difficulty reasoning with and applying knowledge; and who lack the ability to reflect upon their performance and continue the process of learning" (85). Further, our experiences have taught us that integrating experimental approaches into teaching can be extraordinarily difficult, or worse, simply fail to find the institutional support they require (Mehlenbacher 1997).

The project that framed our pedagogical activities across three courses involved our university's electronic "front door" for Web visitors with questions related to outreach, extension, and continuing education. The university's extension site, *Ask NC State,* is poised to play a critical role in disseminating scientific and technical information generated and housed across various colleges, libraries, and extension services. And this site had previously documented, over a series of eight chancellor's retreats in 1997–98, that NC State's extension audiences, in their various capacities as employees, citizens, parents, political officials, and educators, were only partially aware of the immense resources that their land-grant university had to offer them.

Both authors proposed to make usability testing *Ask NC State* a major class project for students from various disciplines enrolled in three different courses:

1. English 583: a special studies graduate course on usability studies (sixteen students)
2. English 517: a graduate course on advanced technical communication (fifteen students)
3. English 421: a Web-based undergraduate course on computer documentation design (twelve students)

Our most immediate goal was to teach students and get them involved in the entire usability testing process of *Ask NC State* by collecting user data about site usage from representative extension audiences, ideally from their hometowns in North Carolina and elsewhere. In this respect, students would act as apprentice usability testers of the Web site, soliciting possible users, designing exploratory tasks, and applying usability-testing data-collection methods in actual contexts of use. Among other usability testing procedures, students were taught to manage small focus groups, to collect talk-aloud audio- and video-recordings of user-interface interactions, to develop matrices for heuristically evaluating Web-based materials, and to analyze and report usability findings and recommendations for an authentic audience, which in this case included sponsors at the Office of Extension, Research, and Outreach at NC State.

Across the three classes, students were taught to prepare test plans, to test materials, to identify and obtain appropriate test subjects, to perform tests, to analyze test results, to generate a test report, to collaborate with others who tested similar subjects, and to prepare a presentation of the results (Rubin 1994).

Benefits of the *Ask NC State* project were to include compiling recommendations and suggestions for *Ask NC State* improvements and revisions, providing students with opportunities to support faculty research or to conduct their own research, and highlighting the innovative *Ask NC State* to students, representative extension users, and Research Triangle–based companies interested in usability practice and theory at NC State. The primary goals of usability testing the Web site, therefore, would be

1. to empirically validate that the Web site's formal features seamlessly meet the needs and support the tasks of its users,
2. to systematically obtain and incorporate user feedback into the Web site's development process, and
3. to report the project findings at a student research symposium held at NC State.

The benefits for the principal faculty members included support from the university extension office to purchase equipment, materials, and supplies necessary to conduct usability testing and some summer research released time for compiling and presenting usability perspectives findings to other units on campus. Importantly, the faculty were not aiming to "profit" from the support that the project received as much as they were aiming to set in place a strong foundation for providing solid research, for supporting some flexibility in instructional decisions, and for sharing usability principles with the large and diversified groups developing Web materials at NC State.

THE EXTENSION CHALLENGE: WHY *ASK NC STATE* VIA THE WEB?

The NC State strategic plan stresses the application of the university's strengths in "graduate education, research, and public service, while strengthening our core mission of undergraduate education" and lists as sites where these core strengths can be applied "classrooms, farms, industries, laboratories, and conference rooms."[1] We contend that Web-based materials designed for extension audiences provide an additional forum that can serve to integrate faculty research, undergraduate and graduate teaching, and outreach and extension goals. As North Carolina State University's Internet "front door" for extension audiences seeking access to NC State's ten colleges, libraries, and numerous extension services, *Ask NC State* is one of twenty home page links accessible from the majority of NC State informational pages.[2] During the week of September 13–20, 1998, *Ask NC State* logged over one million hits per day.[3] Clearly, the need for an evaluation of the Web site's design and usefulness is high.

Moreover, audiences for university Web materials are growing exponentially. The *NSF Indicators Report on Science and Technology, Public Attitudes and Public Understanding* (1998) indicates that Web information "is likely to become a major source of reference-type information in the decades ahead, as access continues to expand."[4] And a recent study by the Angus Reid Group, Toronto, Canada, estimates that the number of Internet users worldwide will increase from the 300 million today to one billion by 2005.[5] We argue that the real and potential audience for *Ask NC State* and other Web-based extension materials can only expand dramatically as well.

In addition to the inevitable growth of information-seeking audiences on the Web, we contend that the usability testing and design of large institutional Web sites have received minimal attention in the research literature. Because *Ask NC State* relies on information from more than ten colleges and institutional offices, even basic goals of consistency, terminology use, and searching become difficult to achieve. In one of the few manuscripts devoted to testing Web sites developed across "agencies, divisions, and departments," Marchionini and Hert (1997) identify two particular challenges:

> The first challenge is that no single person creates such a site—these sites emerge across different departments and eventually are merged under one or a few "home page(s)," but no single individual has full authority over a site or understands everything in the site. . . . The second challenge in large institutional Web sites is an inertia effect. Web sites that get tens or hundreds of thousands of hits per day build a constituency that has invested time in learning navigational and general usage routines and any change will invariably bring comments, requests, and complaints that must be processed in some way, which incurs costs. (1)

Still, we felt that the real-world complexity of the project, combined with the constraints posed by multiple-audience involvement and expectations, offered a rich pedagogical opportunity. Bellotti, Buckingham Shum, MacLean, and Hammond (1995) support this perspective in terms of research in human-computer interaction (HCI), emphasizing that theoretically framed HCI is only achievable when "end-user requirements of the design practitioners are properly understood, and the value of such techniques can be demonstrated" (435). Our immediate goal, therefore, was to explicate the problems we would face in managing this project and to identify task-oriented approaches to addressing them in the space of little more than one semester.

THE PEDAGOGICAL CHALLENGE: IS THERE A "REAL-WORLD" IN THIS CLASSROOM?

Boiarsky and Dobberstein (1998) recommend that documentation-writing classes integrate authentic writing tasks into their syllabi, reminding us that such "assignments are not the usual writing classroom exercises, created 'as if' there were an audience besides the instructor. These assignments require the application of the problem-solving skills . . . in the planning, drafting, designing, testing, and revising of documents that have genuine utility for a broad spectrum of computer users" (45).

But finding ideal sites for research where student activities can contribute to professional developments in the field is not always obvious. Of course, the appeal of using "textbook" assignments is that they are often connected explicitly to the materials covered in class and have well-defined parameters and established standards for evaluation. The solution to the problem posed is frequently readily available or supported by preexisting models. In our experience, attempting to integrate ill-structured "projects" into either undergraduate or graduate classes is exceedingly difficult and something that many instructors learn to avoid. Authentic assignments are difficult to incorporate because faculty find their ill-structuredness difficult to structure, because real-world problems are often messy and therefore appear unfocused, because research does not always produce tidy results easily summarized over the course of a single semester, and because complex problems are more difficult to introduce than well-defined problems (Adams 1993; Øgrim 1991).

But the professional and technical domains that many of our graduating students will enter demand flexibility and resourcefulness on the part of their employees (Denning 1992), and we view this demand as an important reason to mix theory-driven approaches with problem-based or hands-on learning. A major benefit of this instructional approach, according to Kaasbøll (1998), is that

> students work on real-life problems or constructed problems that mimic the complexity of the practical world. In such situations, the students have to think critically through all the information available to sort out the relevant material. When students have the opportunity to define their own problems, they become more involved in their work, and this involvement increases motivation for learning. Because students are assumed to obtain a more profound understanding of the subject area, assessment of problem-based learning should focus more on the students' skills in handling an ill-structured situation than on recalling the textbook. (104)

One area in need of significant research and elaboration is usability testing of Web-based materials, an endeavor that technical communication students are already increasingly involved in as we rush to upload and invent Web-based materials that support our instructional, marketing, information, and administrative organizational needs for distributed documents and support materials (Shneiderman 1998).

Another significant challenge facing the faculty-investigators was the importance of maintaining the intended focus of the three distinct graduate and undergraduate classes involved in the extension project while generating useful data about the *Ask NC State* Web site. This maintenance required ongoing and creative cooperation and flexibility among the instructors. Both English 421 (Computer Documentation Design) and English 517 (Advanced Technical Communication) included usability testing assignments, though neither traditionally required a formal, written report summarizing the results. Students were supposed to learn about the importance of audience feedback in the design of both software and documentation, even though they were not the emphasis of either course. In English 421, students contributed to the class listserv (because the class was completely Web based) and exchanged findings and observations about the Web site via email and attachments. In English 517, students presented their findings during formal oral presentations; students in English 583,however, focused entirely on usability testing issues, and, therefore, the bulk of the data reported to the Extension, Research, and Outreach Office was generated as part of that class.

English 583 students performed various usability tests with four main audience groups using *Ask NC State,* including NC State cooperative extension agents in urban counties, agents in rural counties, job-specific users who can potentially benefit from the Web site in their daily jobs, and members of the general public in North Carolina.

The students performed sixteen tests with nineteen subjects, employing a variety of usability testing methodologies, including contextual inquiry, surveys, questionnaires, interviews, field observations, performance testing, and think-aloud protocols. Usability results discussed in English 421 and English 517 supported the findings of the usability testing class, in addition to adding several new job-specific users to the types of audiences analyzed.

THE RESEARCH CHALLENGE: WHAT IS A "USABLE" INSTITUTIONAL WEB SITE?

We believe that instruction in usability testing provides an opportunity to achieve several highly desirable goals simultaneously for a technical communication curriculum. Because usability testing involves, indeed requires, both theoretical and practical considerations, it provides an excellent forum for accomplishing several practical, pedagogical, and professional outcomes in a single, well-coordinated effort. We also maintain, as Schriver (1997) argues, "usability testing routinely reveals important problems that document designers, even expert ones, may fail to detect" (473).

Though existing research on usability testing highlights improvements to software products (Landauer 1995; Nielsen 1997), few researchers have applied usability testing to large institutional Web sites designed to serve broad extension-based populations. Although usability testing on traditional software applications dates back to the early 1980s (Shneiderman 1998), the usability of Web sites has only recently gained attention at conferences emphasizing usability issues, interaction design, and human-computer interaction (cf. Marchionini and Hert 1997). Research on the results and implications of usability testing on Web sites is therefore still in its early stages. A Web site can be functionally sophisticated and aesthetically appealing without its designers understanding how the Web site is accessed in the context of use and whether the Web site is usable. Functionally, the Web site can be technically innovative and can contain many more features than even required by its users. Aesthetically, the Web site can be visually attractive and graphically creative. However, none of these criteria for success necessarily help Web designers understand what the skills, motivations, and previous experiences of its users might be, what they are attempting to accomplish while accessing the site, how motivated to accomplish the established goals they are, or even in what types of environments they experience user interactions (speeds, machines, system configurations, for example). Linking user knowledge, preferences, and behaviors to desirable Web site "attributes" (such as consistency and layout, navigational support, visibility of features, and relationship with real-world tasks) is a critical goal for all usability performance testing. In addition, Landauer (1995) and Nielsen (1997) have emphasized the substantial benefits of employing even informal usability testing in the process of designing and evaluating software programs in general, and we anticipated that the same benefits would be brought to the *Ask NC State* Web site as well.

A Usability Focus: How Can Multiple Methods Resist Interdisciplinary Solutions?

Implicit in our methodological outlook was the goal of extending classroom-based practices in technical communication beyond the rhetorical range they usually cover. Students, faculty, and administrators learned that contemporary technical communicators can contribute significantly to online design efforts by incorporating techniques from human factors and industrial engineering into our repertoire of strategies for designing usable information. And, importantly, in the wired age, much of the information we prepare is not designed for traditional media but, rather, for online, Web-based distribution. Learning how to analyze the strengths and weaknesses of our online creations is becoming as important a technical communication skill as writing paragraphs or designing usable documents. To expand the scope of what technical communication is and to instruct graduate and undergraduate students in the principles and practices necessary for that expansion, innovative pedagogical approaches and collaborations are essential.

Indeed, the very nature of the field of usability testing makes it a cross-disciplinary enterprise, involving cognitive psychology, software engineering, technical communication, human factors, and sociology. Usability testing *Ask NC State* was a project that, ultimately, required relationships among parties that do not always interact: faculty, students, university Extension, Research, and Outreach personnel, Learning Technology Services (the campus unit charged with moving faculty courses online), the digital library initiatives department (the unit charged with developing online applications for the campus libraries), and information technology (the unit responsible for supporting distributed computing on campus). Moreover, human factors–oriented faculty on campus were housed in disparate departments—from computer science, graphic design, and industrial engineering to technical communication, psychology/ergonomics, and mathematics, science, and technology education. And student chapters of the Human Factors and Ergonomics Society and the Society for Technical Communication, along with their larger regional and national counterparts, represented natural opportunities for further relationships. In this respect, we encouraged students to seek out and share any disciplinary perspectives that might enhance their understanding of the challenge facing any university community attempting to communicate and share information with broader audiences and communities.

Pedagogical Activities

As technical communication faculty committed to developing theory that enhances our intellectual, professional, and disciplinary development, we are also instructors who value and encourage student application of broader principles to real-world information design situations. Schön

(1987) describes an instructional perspective that celebrates reflective practice as "a way of knowing" and argues that

> learning *all* forms of professional artistry depends, at least in part, on conditions similar to those created in the studios and conservatories: freedom to learn by doing in a setting relatively low in risk, with access to coaches who initiate students into the "traditions of the calling" and help them, by "the right kind of telling," to see on their behalf and in their own way what they need most to see. We ought, then, to study the experience of learning by doing and the artistry of good coaching. (17)

Students were charged with identifying design shortcomings of the *Ask NC State* Web site but were not responsible for seeing that an effective redesign occurred or succeeded; that was *our* authentic goal, and, therefore, the usability evaluation and testing methods that we shared with them were methods that we would have practiced outside the instructional context of the three classes.

Various Purposes in Context

The students across the three classes were therefore encouraged to perform tests using a variety of test methods. This encouragement allowed them to tailor the method used to the user group they were testing and to the environment in which they were doing so. It also ensured a large body of data would be collected using diverse usability methods. Agreement in the results of these various methods tends to more strongly support those results (Rubin 1994).

The stated purposes for the tests varied, depending on which audience group a student was testing and what particular types of data the student was seeking. In general, the tests were designed to discover whether people knew about the existence of the *Ask NC State* Web site, whether they could find it on the main NC State Web site, whether they could navigate through it successfully to find specific information, whether they could successfully search it for particular types of information, and whether they understood its navigational structure.

The tests were constructed to provide both performance data and preference data. Performance data show whether users can perform specific tasks with the site and how well and quickly they do so. Preference data indicates users' attitudes toward the site and whether they are likely to return to it.

A Menu of Methodological Choices

Most of the tests included several usability testing methods, with thirteen of the sixteen tests including some type of empirical, performance-based testing to gather performance data. These tests typically included questionnaires and interviews to elicit further data concerning both performance and preferences.

The following list details the methods used. For those not familiar with usability testing conventions, brief definitions for each method are included.

Contextual Inquiry. (A form of field interviewing that focuses on the context of a product's use.) The tester asks detailed questions of users about the product, what they like and do not like about it, what other products they use the product with, problems and errors they experience with the product, and the entire environment surrounding its use, including not only the physical context but also the social, political, and organizational contexts. The method provides primarily preference data.

Survey. (An unstructured interview conducted remotely.) Unlike a questionnaire, the survey is interactive. It is usually conducted by telephone or email. Because it is interactive, the tester can solicit information through open-ended questions. Surveys require more of the tester's time than do questionnaires, but they can yield more valuable results due to their interactive, open-ended nature. Provides preference data and self-reported performance data.

Questionnaire. (A remote, structured interview done on paper or electronically rather than in person.) The questionnaire has a specific list of questions to which users provide answers. Provides preference data.

Interview. (A formal method for gathering data.) Testers prepare a list of questions aimed at providing the type of data needed regarding the usability of the product. Interviews are valuable for gathering information from users that might not surface during lab-based testing, particularly concerning their preferences and attitudes. Provides preference data.

Field observation. (One or more visits to the users' place of work to directly observe them using the product under test.) It affords the opportunity to learn about the users' on-the-job tasks and to see how they use the product in their day-to-day activities. It also allows one to learn about their mental maps for the product. Provides preference data.

Performance testing. (A method to determine how effectively and efficiently users can complete their desired tasks using the product in question.) Provides quantitative, empirical data.

Think-Aloud Protocol. (A method that has test subjects speak aloud as they interact with the product.) It is employed with usability inquiries and performance tests to elicit from the users' statements related to their understanding of the product and its mental mapping, problems with the product interface, and opinions about the usability of the product. Provides primarily preference data, but can supplement and enhance performance data.

The tests yielded a large body of data, concerning both user performance and preferences. Overall, the results indicated that users had highly positive opinions of the site's intended purpose, but that they were frustrated by the mechanics of trying to use it successfully. In general, the results indicated low performance levels in finding the site, understanding its relationship to the NC State extension service, navigating within the site, and searching for and finding specific information. The preference data generally showed that users were confused about the site's purpose and operation, but that they found information to be very useful once they figured out how to get to it.

The students each completed a detailed usability report explaining the results of their tests. They then worked in groups, with other students who had tested the same audience, to compile a summary presentation of the overall results to be given to the NC State extension service. The composite report was presented in a combined meeting of the three project classes, the NC State student chapter of the Society for Technical Communication, and personnel from the Office of Extension, Research, and Outreach. Further, a written composite report was presented to the extension office (Dicks and Mehlenbacher 1999).

Pedagogically, the assignment required students to duplicate conditions often encountered in workplace usability testing. They had to perform their tests individually, not uncommon for technical communicators. They had to collaborate with groups to create audience-specific reports and composite reports. They learned about the importance and the social and political implications of how usability information is reported to a client. They further honed their knowledge and skills in dealing with a complex rhetorical situation and in reporting information both orally and in writing.

Their final report concludes with a number of recommendations for improving the *Ask NC State* Web site, including changing the name, making the site's purposes and functions clearer, improving navigational organization, repairing broken and outdated links, and repairing the search engine.

The Development of Usability Principles

One of the outcomes of teaching reflective practice is the development of general principles for guiding future research and, ultimately, for folding back into future instructional practice. In this respect, our experience working with students on an authentic problem—the evaluation of an operational university Web site—helped us to further develop our understanding of usability issues related specifically to Web site design. The appendix summarizes usability principles for Web site design that evolved during the course of the project and that we continue to refine and extend. Though not the focus of this chapter, we viewed the development of these principles to be explicit evidence that teaching, research, and extension activities—frequently separated historically in university environments—can feed into each other in creative and exciting ways.

Another outcome of the experience was that several students from the classes joined other usability-related efforts on and off campus following the semester and, in this way, contributed to design processes as well as product development. One graduate student presented with the authors on instructional usability and student learning at NC State's 1999 Summer Institute for Distance Learning, and she later joined the university's Learning Technologies Service as a full-time employee.

IMPLICATIONS AND NEW DIRECTIONS

When assessing how successful or unsuccessful any long-term project has been, it is tempting to recount the positive and to de-emphasize the unresolved problems encountered in process. We know that students were excited by the authentic problem-solving situation and that their end-of-semester presentations revealed thoughtful and professional engagement in the overall goals of the project.[6] But we are also aware of the proviso about teaching effectiveness that Almstrum et al. (1996) provide:

> 'If I'm satisfied and my students are satisfied, have I done a good job?' Yes, if 'goodness' in teaching is simply a matter of mutual satisfaction. No, if 'goodness' has something to do with learning, unless we establish that mutual satisfaction is a reliable indicator of learning.

Unfortunately, mutual satisfaction in instructional situations is not always correlated with learning performance (Kaasbøll 1998). Sometimes, learning can be exhausting, time consuming, difficult; it may even involve dramatic cognitive dissonance on the part of engaged students. Our experience attempting to integrate research, teaching, and service work, though rewarding, was not without challenges.

First, because the authors have continued their involvement in the institutional challenge of integrating usability methods into the development of Web-based materials at NC State, we are aware that concluding the students' "experience" with usability testing at the end of the semester may have misrepresented the long-term complexity of the problem. That is, for our three classes of students, the problem of identifying and analyzing potential audiences for the *Ask NC State* Web site ended when they presented their findings to their class and the shareholders with the Office of Extension, Research, and Outreach. But the thornier *problem* of creating a usable Web site for audiences external to NC State is far from over, and, in fact, the authors' roles in that process continue to this day.

Of course, many real-world problems are by nature complex and do not operate in isolation; instead, they touch on and influence other issues, political, cognitive, interpersonal, and institutional in nature (Spiro, Vispoel, et al. 1987). Moreover, Spiro, Feltovich, Coulson, and Anderson (1989) have shown that simplifying complex concepts may actually lead to erroneous interpretation on the part of students. We attempted to avoid oversimplifying the challenge of collecting data and reporting it to a motivated and interested audience. In particular, we reminded students that an early indicator of larger success would be seeing their recommendations integrated in subsequent design efforts of the Web site, rather than being congratulated for clearly presenting recommendations for redesign following their data-collection efforts.

As well, we were constantly mindful of the nontraditional role we were asking students to play in having them serve as apprentices to faculty researchers. Unfortunately, many instructors compelled to describe the engaging aspects of using authentic projects in the classroom do not acknowledge the possibility for conflict of interest. That is, by asking students to collect real data and to contribute products that will be used in actual corporate and academic environments, are faculty-researchers possibly guilty of coercing students to volunteer their services for the currency of course grades? And, if so, what are the students' rights in such instructional situations? Can students choose to withdraw from particular project assignments if they feel their values are being violated? In the case of a Web site aimed at providing useful information about a state university to the general public, the situation is perhaps less open to criticism, but what about the design of a Web site for Westinghouse or a brochure for a local abortion clinic? We addressed these concerns by raising them as explicit topics and by making certain that the *Ask NC State* Web site project did not dominate the syllabi of the three classes. And we did not feel that we were entirely alone in the university community in having students engage in activities that had results the university benefits from: most psychology departments routinely require students to act as subjects in psychology experiments for course credit, and many independent study and internship programs are framed by an exchange between student labor and student learning and training. Technical communication programs need to be particularly careful to periodically address their position on issues of "sponsored" research and instruction versus theoretical isolation from the professional world around them.

Despite the challenges we have raised here, our goal of combining usability testing with instruction and research has yielded valuable results for the students, the faculty-researchers, and for the university. The students sharpened their skills and knowledge of usability testing methods. They also learned about the difficulties of designing online information in a complex rhetorical domain. The faculty-researchers developed numerous relationships with other campus researchers and groups interested in usability studies. Further, they reaffirmed the value of employing diverse usability methods for testing a multipurpose communication medium. The university has benefited from an increased level of activity and understanding regarding the importance of performing usability tests on its communications with its constituents. It will also benefit from initiating a process to improve the quality of the Web site it uses for offering extension services to help fulfill its role as a land-grant institution.

ACKNOWLEDGMENTS

We are grateful to many people who helped us through the course of this project. They include, especially, Everette Prosise, assistant vice-chancellor, and June Brotherton, former vice-chancellor of North Carolina State's Office of Extension, Research, and Outreach, Harry Nicholos, Webmaster with the Office of Information Technology, and Caroline Beebe, director of North Carolina State's Digital Libraries Initiative Department. In addition, the authors wish to thank the forty-three students in our English 583, 517, and 421 classes, who enthusiastically participated in the experimental *Ask NC State* project. Without the energy and commitment of these students, none of the issues and outcomes discussed in this chapter would have been possible.

APPENDIX

EXTENDING USABILITY RESEARCH

Usability Principles for Web Site Design

(cf. Bevan 1998; Nielsen 1994, 1997; Selber, Johnson-Eilola, and Mehlenbacher 1997)

Accessibility	Has the website been viewed on different platforms, browsers, modem speeds? Is the site ADA compliant? Have ISO-9000 standards been considered?
Aesthetic appeal	Does the screen design appear minimalist (uncluttered, readable, memorable)? Are graphics or colors employed aesthetically? Are distractions minimized (such as movement, blinking, scrolling, animation, and so on)?
Authority and authenticity	Does the site establish a serious tone or presence? Are users reminded of the security and privacy of the site? Are humor or anthropomorphic expressions used minimally? Is direction given for further assistance if necessary?
Completeness	Are levels clear and explicit about the "end" or parameters of the site? Are there different "levels" of use and, if so, are they clearly distinguishable?
Consistency and layout	Does every screen display begin with a title/subject heading that describes contents? Is there a consistent icon design and graphic display across screens? Are layout, font choice, terminology use, color, and positioning of items the same throughout the site?
Customizability and maintainability	Does printing of the screen(s) require special configuration to optimize presentation, and, if so, is this indicated on the site? Are individual preferences/sections clearly distinguishable from one another? Is manipulation of the presentation possible and easy to achieve?
Error support and feedback	When users select something, does it differentiate itself from other unselected items? Do menu instructions, prompts, and error messages appear in the same place on each screen?
Examples and case studies	Are examples, demonstrations, or case studies of user experiences available to facilitate product learning? Are the examples divided into meaningful sections (such as overview, demonstration, explanation, and so on)?
Help and support documentation	Does the site support task-oriented help, tutorials, and reference documentation? Is help easy to locate and access on the site? Is the help table of contents or menu organized functionally, according to user tasks?

Intimacy and presence	Is an overall tone that is present, active, and engaging established? Does the site act as a learning environment for users, not simply as a warehouse of unrelated links?
Metaphors and maps	Does the site use an easily recognizable metaphor that helps users identify tools in relation to each other, their state in the system, and options available to them?
Navigability and user movement	Does the site clearly separate navigation from content? How many levels down can users traverse and, if more than three, is it clear that returning to their initial state is possible with a single selection? Can users see where they are in the overall site at all times? Do the locations of navigational elements remain consistent? Is the need to scroll minimized across screens and frames within screens?
Organization and information relevance	Is a site map available? Is the overall organization of the site clear from the majority of screens? Are primary options emphasized over secondary ones?
Readability and quality of writing	Is the text in active voice and concisely written (5–14 words per sentence)? Are terms consistently plural, verb + object or noun + verb, and so forth, avoiding unnecessarily redundant words? Do field labels reside on the right of the fields they are closely related to? Does white space highlight a modular text design that separates information chunks from each other? Are bold and color texts used sparingly to identify important text (limiting use of all capitals and italics to improve readability)?
Relationship with real-world tasks	Is terminology meaningful, concrete, and familiar to the target audience? Do related and interdependent functions appear on the same screen? Is sequencing used naturally, if sequences of common events are expected?
Reliability and functionality	Do all the menus, icons, links, and opening windows work predictably across platforms?
Typographic cues and structuring	Does text employ meaningful discourse cues, modularization, chunking? Is information structured by meaningful labeling, bulleted lists, or iconic markers? Are legible fonts and colors employed? Is the principle of left-to- right placement linked to most-important to least-important information?
User control, error tolerance, and flexibility	Are users allowed to undo or redo previous actions? Can users cancel an operation in progress without receiving an error message? Are multiple windows employed, and, if so, can they be manipulated easily?
Visibility of features and self-description	Are prompts, cues, and messages placed where users will be looking on the screen? Do text areas have "breathing space" around them? Is white space used to create symmetry and to lead the eye in the appropriate direction?

14

AT THE NEXUS OF THEORY AND PRACTICE

Guided, Critical Reflection for Learning Beyond the Classroom in Technical Communication

Craig Hansen

Other chapters in this collection (Grabill or Dubinsky, for example) invite technical communication educators to enrich both the curriculum and the students with service-based learning experiences. This chapter builds on those ideas and extends them in a significant way: I advocate for integrating a reflective element grounded in critical thinking, suggesting how this integration can bridge theory and practice for students in a vibrant way.[1]

Such connections are important—and perhaps essential in a volume that promotes innovative pedagogy. Locating the balance of theory and practice has been one of the most persistent controversies in technical communication as an academic discipline. Educators want to instill students with theoretical perspectives that build critical thinking and decision-making skills and that maintain flexibility and perspective over time. But educators also feel pressure to prepare students with applied, practical skills to ensure immediate, successful entry into the workplace. These can be competing goals and sometimes divide the communities of practitioners and academics. Finding balance becomes especially relevant in a current national culture that places increasing emphasis on work preparedness and accountability for higher education

But before I suggest applications for critical reflection (including several sample cases), I would like to broaden the discussion with regard to learning that takes technical communication students beyond the classroom. To this end, I include a brief description of two additional learning experiences—client-based learning and internships.

CLIENT-BASED LEARNING AND INTERNSHIPS

Client-based learning means that students undertake projects (usually in groups) with "real" clients (commercial or nonprofit) external to the classroom. The projects might be product oriented, in which students create manuals, Web pages, brochures, or proposals, or they might be process oriented, in which students act as communication consultants to help solve problems or recommend new procedures. Whatever their form, these projects are part of assigned and evaluated class work. Client-based projects afford students valuable workplace contact and can help students at all levels (Wickliff 1997) encounter and internalize genres of workplace writing (Blakeslee 2001).

Internships differ from client-based projects in two ways: 1) in most cases, no formal content instruction takes place: the learning focuses on performance of tasks for an external site—it's an immersion experience; 2) in many cases, these are solitary learning experiences because students typically have more opportunities to take part in individual rather than group internships.

It is interesting to note that studies (not within technical communication) have linked internships to improved employability and starting salary (English and Koeppen 1993) and, in addition, to improved academic performance following the internship (Knouse, Tanner, and Harris 1999). Further, internships can provide an effective transition into the workplace for students from disadvantaged or marginalized backgrounds (Cates-Melver 1999; Mellander and Mellander 1998).

Both experiences clearly benefit technical communication students (hence their popularity with technical communication degree programs). In addition to the benefits previously mentioned, I would add that because these projects are never just writing projects, students develop a recognition of the relatedness of multiple literacies (oral, written, visual, and technological) and the complexities of communication within organizations (see, for example, Tovey 2001; McEachern 2001).

Service- and client-based projects and internships are not the only options for learning beyond the classroom. Other scenarios that work well to acquaint students with workplace values and challenges (and gain value from the reflective practices described later) include cooperative agreements common in engineering programs (see, for example, Wojahn 2001) and mentoring relationships (Kryder 1999).

In any of these learning experiences, students encounter the complex social and political contexts that surround the development of information products. And for more immersive experiences, students are necessarily participants in these contexts. All this immersion creates an opportunity for deeper, multilevel learning and guided, critical reflection in ways that can play a key role in that learning process.

ENCOURAGING CRITICAL PERSPECTIVES

Importance of Reflection

When students venture beyond the traditional classroom, they often discover that the work of creating information products for real clients is far messier than the rule-driven advice in many textbooks (for more observations on the relationship between theory and textbooks, see Herndl 1996a). Students can become distracted by the complexity of the experience and struggle to maintain a learning focus. The instructor or faculty advisor can help with this focus by connecting students to useful perspectives about the experience (perspectives I would argue should be critical and theory based). The key to perspective is reflection. In fact, it might be argued that to make any of these experiences a form of *academic* learning, students need time and structure for reflection (Watson 1992). When students are self-aware, they tend to be more exploratory, more receptive to new ideas, and simply gain more insight into the experience. They may also be more willing to be critical (and perhaps even resistant) toward organizational culture and the power structures they encounter—the "dark side of the force" (155) as Carl Herndl (1996a) describes it. I agree with Herndl that this resistance should be an important part of technical communication pedagogy, and in the more practical advice on guided reflection described in this chapter, I point to some ways instructors can encourage students to consider these possibilities.

Creating Meaningful Reflection

Meaningful reflection can take many forms and can take place at any point during the learning experience—and in fact should occur throughout it—with perhaps a special effort toward summative reflection at the end. Written reflection is especially important because of the reinforcing links among thinking, learning, and writing. For service- and client-based learning, students can reflect individually through journals or assigned writings; as groups, they might formally reflect on their experience in, among other possibilities, a summative piece that accompanies the final project. For internships, a journal is important, as is a final summative paper and, as suggested later, an internship seminar.

To avoid journals or other reflective writing that simply record a descriptive chronology of activities, instructors can provide prompts that encourage informed, critical thinking. These prompts can be part of formal assignments or be more informal suggestions. Guided reflection might include the following possible questions.

General questions:

What classes prepared you for this experience?

What previous life experience prepared you?

What should you have known before you went into it?

What are your initial impressions about the client or site?

What has surprised you about the experience?

How did this project/internship fit into your personal or civic goals? Your value system?

Would you recommend this experience to another student? Why or why not?

What kinds of problems did you encounter? How were you able to solve them or work around them?

What types of communication skills are important to be successful in this internship?

What are some key things you've learned from this experience?

Questions specifically for service- or client-based learning:

What is the goal of your project? When it's all done, what should it accomplish?

How did your group divide the work? Was it successful? How would you do it differently?

How did your project change over time, in terms of expectations or schedule? Why do you think this happened?

What do you think about your relationship with the client? What were the challenges? What seemed to go well?

How would you describe your client's workplace?

Questions specifically for internships:

Describe the hierarchy of people around you. How long did it take you to figure it out?

What forms of communication are used at your site, and how do you think they influence the organizational culture?

What seems to work particularly well at this site? What seems problematic to you?

Were there any mentors for you, and how did they work with you?

Do you think your contributions are valued? Why or why not?

What would you change about this site if you could?

How do people work together? Are there teams? How do they divide up work?

Do you now have any concerns or observations about work-related issues such as telecommuting, work/home distinctions, physical commuting, others?

Written reflection encourages introspection by individuals and small groups. It may smooth the course of the project or internship, contribute to a higher quality product, and provide a useful record of thoughts and activities. But written reflection still has limitations and is not the only option for meaningful reflection. Students can also gain a great deal of insight by interacting with each other and the instructor. To this end, I would argue for regular large group discussions, guided by an instructor as a complement to both service- and client-based learning as well as internships.

Creating meaningful discussion seems fairly straightforward—even unavoidable—with many service- and client-based projects, as they exist within the structure of a class, and students find these projects engaging. Yet, this type of interaction is probably more rare with internships. Internships can be solitary, isolating learning experiences. Students frequently have little status at their sites and may have little contact with other students. An internship seminar significantly reduces this isolation. Here, interns from a variety of sites meet regularly with a faculty advisor to discuss their impressions and concerns. A seminar also provides an opportunity for an instructor to introduce critical perspectives.[2]

Bringing Theory into the Reflection: Sample Cases

Once students are involved in a well-structured service- or client-based project or internship and have participated in ongoing reflection, they are prepared for more critical analyses of their experiences. The instructor is vital in this process. Probably, the easiest way to illustrate the role of the instructor—and for that matter, the value of introducing more theoretical perspectives—is to look at several sample "cases" that represent common student comments and reactions to these experiences. (I have drawn these cases from real student experiences.) Each case suggests how an instructor might steer somewhat vague observations (or complaints) into focused (and perhaps revealing) critical thinking. In these cases, I make no attempt to provide a detailed review of the various theoretical perspectives: given the variability of these experiences and the differing interests of instructors, such an attempt would be well beyond the scope of this chapter. Rather, what I provide is more analogous to signposts—directions that invite further exploration.

It is also important to note that discussions aimed at enhancing awareness and critical thinking may not actually solve problems that students may be experiencing. Instead, I think reflective practice provides students with new perspectives. Although it is not always possible (or even desirable) to "operationalize" some of the more abstract discussions suggested later, students may be able to use insights to alter their communication or project strategies.

For each case described, I present a brief narrative and a series of questions that can serve as general prompts for critical reflection. These activities are followed by a discussion relating the prompts to a case and some suggestions for additional resources.

CASE ONE: "I'M CONFUSED": ADAPTING TO THE WORK ENVIRONMENT

Description

A group of students in a software documentation class have undertaken a client-based project. Their client, a small technology start-up, has developed a software product designed to facilitate small business accounting. At this point, the students have met with one representative of the company and have had a week to familiarize themselves with a prototype product. They've just had a second meeting that involved more people from the company, including midlevel managers from both the development and marketing areas of the company. What seemed like a clear-cut documentation project now seems complicated. The marketing and development people disagreed on the target audience for the product and even on the functionality of the product itself. During class after the second meeting, the students express confusion.

Prompts and Discussion

A common response from students in the early stages of a service- or client-based project (or an internship) is a certain amount of confusion. Often the project seems clearest at the very beginning. But once students start working in earnest with clients to pin down audience, purpose, research requirements, design goals, and other matters of substance, a certain amount of disorientation and frustration sets in. When frustration happens, it is useful to engage the large group of students—not just those expressing confusion—in a discussion that might take any or all of the following directions.

Culture and Community

Perhaps the confusion arises from misunderstandings, particularly about each other's priorities. How might students describe their impressions of the culture at the site? What are the commonalities that bind the people in this workplace together? Can they speculate about their values? Their priorities? Where might students' values and priorities overlap? Where do they differ? (See, for example, Deal and Kennedy's classic text, *Corporate Cultures: The Rites and Rituals of Corporate Life,* 1982.) Drawing on qualitative or ethnographic research methodology, how might students try to understand local culture? How might they see their roles as participants/observers? What might be problematic about that view? (See, for example, Campbell 1999; Herndl 1991.)

In this particular case, the students initially had little feel for the culture of the company. But a few conversations with employees revealed that, until recently, this company had been a software consulting group, and its attempts to market its own products was relatively new. The development side seemed to resent the rapid growth of the marketing side. In individual writings and class discussion, it became clear that the students identified with the software developers. With that awareness, and, as a result of a discussion that continued over several class periods, the students developed a strategy that basically followed the software developer's interpretation of the product and the marketing department's interpretation of the target market. In this case, the "other" was the marketing department, and students had to learn to trust marketing's expertise, even when they didn't share all of its values.

Power and Hierarchy, Gender

From a related point of view, perhaps the difficulty arises out of the relative power positions of the students and the clients. How might individual people at the work site view this new, short-term relationship? What is the power hierarchy at the site, both formal and informal? Where do students or interns fit in this hierarchy? How does one move around or within the hierarchy? How might gender play a role, particularly in perceptions of authority or power? How are power hierarchies created and maintained? (See, for example, Flynn 1997, on feminist theory/gender issues; Baker and David 1994, on power issues.)

The students found issues of power particularly engaging in this case. Part of the reason that the students "liked" the software developers was that they seemed to appreciate the effort involved with producing effective documentation. The marketing manager seemed comparatively dismissive. The students gravitated toward the area where they had most status. Also, the students found it difficult to identify the balance of power between the two departments. Despite class discussion on this issue, as well as some efforts at the site to clarify it, the students never did determine clearly who had the final say in decisions regarding their documentation project: thus, they proceeded with the two-pronged strategy described previously. The students, as a group and during class discussions, did not feel that gender played a significant role in this particular project.

Discourse and Models of Communication

The theme of culture brings up another question: Is the source of the confusion actually language? Do they—the "other"—actually know what the students are saying? How does the concept of discourse community apply here? Are communities and cultures defined by their discourse, or do they define their cultures (thereby creating insiders and outsiders) with language? Do communication models help shed some light? In terms of a Bakhtinian dialogic, is there an understanding that follows the utterance? Or what is the interference here between sender and receiver? How might students know if somebody actually understands? (For an excellent discussion of all these issues, see Gregory Clark's *Dialogue, Dialectic, and Conversation: A Social Perspective on the Function of Writing,* 1990.)

The students in this case, who purposefully chose to work mostly with the software developers, did not feel there was a significant gap in communication, either in terms of style or discourse. A class discussion about the students' choice to stay on relatively familiar turf was particularly interesting: their alliance with the software developers became an actual choice (whether it was wise or not is debatable), rather than an unexamined drift. In "The Overruled Dust Mite: Preparing Students to Interact with Clients," Lee-Ann Kastman Breuch (2001) directly addresses this issue—by presenting a case where students, pleased with their design approach, simply do not hear the direct wishes of their client—

and discusses the causes and complications involved in communication breakdown. This article might be good preparation for students beginning projects beyond the classroom (especially for students in technical communication degree programs).

CASE 2: "THEY KEEP CHANGING THEIR MINDS": DEALING WITH SHIFTS AND SLIPS

Description

Students in a document design class have undertaken a project somewhere between service- and client-based learning. They are working with a county sheriff's department to create a "handout" for crime victims. In the past, the sheriff's department has handled this information informally, relying on officers, public agencies, or department counselors to acquaint victims with their rights. Now, they would like to enhance consistency with a document. The students quickly determine that a one-page handout will not be sufficient and begin design work on a compact brochure. As work progresses, word spreads about the project through the sheriff's department and among relevant agencies and nonprofit groups. The students' liaison with the department passes on the numerous requests for enhancements. As the weeks progress, the brochure has become a small book, with chapters for victims of different types of crimes, and the focus begins to shift from informing victims to concatenating department information. As the end of the semester approaches, the students realize their project is perhaps hopelessly far from completion.

Prompts and Discussion

Nearly all communication projects experience shifts in standards, content, schedules, budgets, personnel, and expectations. These changes can be frustrating for interns and occasionally catastrophic for students working on service- and client-based projects, as they work within the time strictures of a finite class. These strictures can become complicated (but not impossible) difficulties to resolve. In these situations, guided reflection might focus on any of the following categories.

Constructive Processes

Perhaps change results from the interaction of larger forces. Why do people change content or design? Who are the stakeholders for the resulting product? Do they have the same interests? How does this project encourage (or even force) negotiation among the stakeholders? How does the process of negotiation work? How long does it take? What other factors affect it? Given all these constructive processes, do communications professionals record, refine, or invent truth? (Scholarship on social construction is helpful here; see, for example, Bruffee's [1986] classic essay, "Social Construction, Language, and the Authority of Knowledge" or the essays in Blyler and Thralls's [1993] *Professional Communication: The Social Perspective.*)

What is the larger context of this project (for instance, industry, nonprofit sector)? Are there other information products like it? Do these serve as models or something to be avoided?

In this case, the students encountered a latent and undiagnosed communication vacuum. In the written reflection and in the small group and class discussions, students displayed an initial enthusiasm for their project that, over time, became disillusionment and even fear. They did not want to do this poorly, and the increasing, shifting demands for the project seemed to remove it from their control. Discussion helped the students diagnose the situation: the students determined that the project had caused the sheriff's department and other agencies to really think this process through for the first time. They did not have any consensus or models to present to the students; rather, the students, with the clock ticking, had become observers of a vital constructive process—a situation familiar to professional communicators.

Intertextuality, Multiple Literacies

Perhaps students can see these changes as related to a natural development of ideas over time. Why might ideas change over time? Where did the ideas for this project originate? What actual forms have they likely gone through (from hallway conversations to email to phone calls to memos to presentations to reports and so forth.)? Are these ideas likely to continue to appear in different media, in different forms over time, even after the project is completed? What types of literacies might people require to track this idea over time (such as oral, written, visual, or technological)? How might professionals know when to freeze change, at least for the purposes of completing one information product? (For a study on intertextuality in a corporate setting, see Hansen 1995.)

Here, the students have encountered a fragmented communication strategy. Crime victims gained information from different people in different ways. As the sheriff's department moved toward some level of consistency for the sake of the project, they uncovered many existing inconsistencies—and even some reluctance to move toward consistency. In class discussions, the students noted that, as the means of communication shifted from primarily oral to written, the sheriff's department seemed to want to add information for every possible contingency. The students realized that, in the past, crime victims received information on a case-by-case basis, with the information varying according to individual circumstances. The brochure, as a static document, really required a different approach: here, shifting modes of communication required a fundamental shift in communication strategy. (This shift was another issue that slowed the project.) As a result, the students worked with the department liaison to identify information that might be common to all crime victims: even this proved problematic.

In the end, the students could not create a brochure that met the original goals. Realizing they were facing a dynamic information situation—and running out of time—they recast the project and produced a highly useful brochure for victims of property crimes.

CASE 3: "I DON'T LIKE MY CLIENT OR MY GROUP MEMBERS": CONFLICT AND COLLABORATION

Description

In a Web design class, a small group of students are creating a "community bulletin board" Web site for a neighborhood association. The association represents a diverse urban neighborhood, one that has made great strides in developing a new sense of community. They view the Web site as an essential communication vehicle both for the residents of the neighborhood and for external audiences. The student group undertaking this project is talented. Two members have significant professional experience in Web site development. Rather quickly, however, the instructor notices that the group seems to be falling behind other groups in the class (who have similar projects). While sitting with the group, the instructor observes long silences, stiff body language, and a general lack of engagement. Group members insist they have no problems within the group. Several weeks into the project, a representative from the neighborhood association contacts the instructor and voices concern about the progress of the project and the responsiveness of the student team.

Prompts and Discussion

Some level of conflict is inevitable in technical communication projects beyond the classroom, simply because these projects require high levels of coordination, real deadlines, and strong personal attachments to ideas, designs, and written work. This conflict may well turn out to be the most challenging area for the instructor. And it can be delicate. If the conflict lies primarily within a student project team, the instructor can intervene in various ways, from quick encouragement to reorganizing or even dismantling the group. However, if conflict forms between the students and site, the instructor has less control, sometimes fewer options, and many more variables to manage. It may be best with external conflict for the instructor to act as an intermediary between the site and students—or, in some situations, as the students' advocate. (Given the potential liability issues, an instructor should not hesitate to terminate a project if relationships really go awry.) However, through large group discussion, students can gain insight into collaboration that may serve them well in the future.

(Re)organizing the Project Team—Small Group Dynamics

What are some ways to organize a project team (strong leader, democracy, specialized functions, subteams, and so on.)? How might different projects fit best with certain ways to organize? What roles might individuals assume within a small group? What effect might individual team members' skills have on team organization? Does a team have to keep the same organization throughout an entire project? Why might a team reorganize during different phases of a project?

In this case, the instructor decided, even though most groups were functioning well, to spend some extra time reviewing small group dynamics and models for team organization. After class, some members of the student team approached the instructor and asked for help. They noted that the two students with the most experience constantly disagreed and that this disagreement was not only slowing their progress but also effectively silencing the other students. The instructor met with the students and reviewed their team organization. One experienced member had assumed a role as a central coordinator, a "boss," without a real mandate from the rest of the group. The instructor asked the students to work from the material on small group dynamics and develop an alternative structure within the group. This they did, dividing the project into discrete steps and assigning responsibility for each stage to a different student.

Conflict

Much has been written in technical communication scholarship about conflict and collaboration. It is probably one of the stronger areas of the field. Many resources exist, but what I have found most useful for this immediate purpose derives from one of Rebecca Burnett's (1993) many works in this area, "Conflict in Collaborative Decision-Making." It is useful to ask students to focus on the nature of the conflict:

- Is the conflict based on how to do things (procedural conflict)? How long has this gone on? At what point does this cease being a useful type of conflict? How might people get past this type of conflict?
- Is the conflict based on personalities (affective conflict)? Is this an appropriate type of conflict for a professional project? Is this a common type of conflict? Is it a productive or useful type of conflict? How might people get past this type of conflict?
- Is the conflict based on real issues with the project (substantive conflict)? In other words, is the conflict over the content of the project, its design, strategies, and such? Why is this a more useful type of conflict? What might be a negative consequence of no substantive conflict? Can this develop into other types of conflict? What is the best way to manage substantive conflict?

Here, the students were able to determine that they were mired in procedural conflict (with a touch of affective conflict). The battle for power between the two experienced members manifested itself in continual disagreement about how to proceed with the initial stages of the project. This simple diagnosis was useful for students, and they were able to move beyond it once they reorganized.

The conflict that began to develop with the association was a little more problematic. The association representative had high expectations for the group (perhaps unrealistically high) and, with the group's initial inability to move forward, had all but dismissed the group as developer of their Web site. It became part of the group's project to rebuild this trust through real progress and better responsiveness. (And it was also a learning experience for the instructor, to establish carefully defined expectations between sites and students.)

CASE 4: "I DON'T LIKE WHAT'S GOING ON HERE": ETHICS AND SOCIAL RESPONSIBILITY

Description

A technical communication student has become an intern for a large annual arts festival. The internship description defines the intern's responsibilities as primarily to "aid in the design and creation of promotional materials." One of these projects is the "Director's Update," a newsletter prepared for festival participants and sponsors. The intern works directly with the director to produce the newsletter. The director compliments the intern's work and asks the intern to take on additional communication-related projects for the director. The intern is flattered and enjoys preparing press releases and presentations. But the responsibilities begin to shift, and the intern finds herself making appearances at meetings and even press conferences in the director's stead. More troubling, the director asks the intern to slightly misrepresent information—about the director's activities, about some details of the festival organization. The level of misrepresentation seems relatively minor, but the intern is distinctly uncomfortable and confused by the course of the internship. She approaches the faculty internship advisor for help.

Prompts and Discussion

Less frequently than in the other "cases," a student may be dissatisfied with a site's (or fellow student's) general approach to the project, to the handling of specific situation, or to treatment of an individual. These types of issues—when they are not better described as conflict or quickly handled as potential liability problems—require an exploration of ethics and social responsibility. This exploration may be intensely personal for students, as they compare personal values with those of organizations, individuals within the organizations, or other students. It may not be advisable to make these individual situations part of general group discussion, but addressing these significant issues is important.

Much of the groundwork for a discussion of ethics and social responsibility might originate from a conversation about culture, such as that suggested in Case One. Beyond that, however, an instructor and student might explore some of the following issues.

What are "ethics"? Do professions have standards for ethical behavior (for example, the Society for Technical Communication)? What happens when individuals find themselves in situations where they are asked to violate either personal or professional ethics? Or to stand by and watch others act in ways that appear unethical? What is the duty of the individual to act? What are the options for action? This area has also received a good deal of attention in technical communication (see, for example, Faber 1999; Markel 1997; and the *Technical Communication Quarterly* special issue on ethics, volume 10, issue 3, 2001).

A further goal here might be to show that a sense of social responsibility is not simply (or only) an artifact of the liberal university environment: What do we mean by social responsibility? To whom is it owed? Why is it owed? Is it optional? How is social responsibility part of personal value systems? Is it also part of ethics? In this regard, Donna Kienzler's (2001) article, "Ethics, Critical Thinking, and Professional Communication Pedagogy," provides useful guidelines. She combines critical thinking and ethics and asks students to evaluate situations by applying four related activities: identifying and questioning assumptions, seeking input from diverse voices, connecting with the relevant communities, and becoming actively involved. This approach helps students clarify the problem as well as understand their values and sense of responsibility.

In this case, after some investigation and negotiation, the faculty advisor removed the intern from the site (and completed the internship with related activities within the university). But the advisor also asked the student to write reflectively about her experience with the internship, using some of the questions listed previously as prompts. What bothered the student most about this situation was finding herself in a position where she was "in over her head," where she was not sure about boundaries and consequences for her actions. This misrepresentation of information violated her own values, but she was more troubled by how she had felt personally compelled to follow the directives of her site supervisor. She wanted to do a good job and meet expectations—and she appreciated the praise from the director. This experience caused her to reflect, in a way Herndl (1996a) describes, on the power of organizational hierarchy and convention to subvert personal responsibility. Reflection allowed the student to both explore her discomfort with the situation and broaden the experience into observations on the nature of compulsion and resistance.

CONCLUSION

In general, guided, critical, written reflection and discussion invite students to process these project and internship experiences with an eye toward

- generalizing their learning, so that they can turn these specific experiences into frameworks for understanding future experiences,
- connecting their learning to larger ideas—in essence, seeing how a theoretical lens (whether feminist, rhetorical, social, cognitive, or one of many others) sheds light on real experiences—and
- appreciating the centrality of communication in human culture—in all its forms and complexity, in all kinds of workplaces and all types of work.

Service- and client-based projects and internships provide many practical learning opportunities for students. But I would argue that they can provide even more: when combined with opportunities for guided reflection, students can gain a deeper understanding of the theoretical basis of technical communication, which in turn might allow them to maintain and sharpen a critical, thoughtful perspective about their lifelong work experience. In this way, service- and client-based learning and internships can foster significant intellectual growth.

Currently, the Internet offers significant potential for service- and client-based projects and internships that might be global in scope, greatly enhancing the cultural learning integral to these experiences. But this enhancement will also pose interesting new challenges. As the Internet (or a successor technology) becomes the workplace of the future and as geopolitical boundaries fade in professional communication, learning opportunities that give students critical, self-aware experience with issues like culture, power, and ethics will become only more important.

15

(RE)CONNECTING THEORY AND PRACTICE

Academics Collaborating with Workplace Professionals—the NIU/Chicago Chapter STC Institute for Professional Development

Christine Abbott

In this chapter, the type of "innovation" described is neither earth shaking nor futuristically high tech; rather, it captures the original meaning of innovate, the etymology of which is from the Latin *innovare,* (to renew): *in* + *novare,* to make new, renew. Specifically, it describes how the technical communication program at Northern Illinois University has been renewed and enriched by its affiliation with the Chicago Chapter STC Institute for Professional Development. Through the collaborative efforts of many teachers, researchers, and practitioners of technical communication, the Institute offers courses for traditional students and working professionals that merge theory and practice and that carry either graduate or undergraduate credit or continuing education units (CEUs). Among the features that make this partnership unique is the extraordinary amount of interaction that occurs at every stage of the courses—before, during, and after—and at every level, from the boardroom to the classroom, and in every way, from curricular planning, marketing, teaching, and mentoring, to postcourse follow-up assessments and focus groups.

INTRODUCTION: THE CHALLENGES

Technical communication programs face a number of challenges now and in the future: shrinking budgets, downsizing of faculty, increased competition for student market share, growing territoriality among departments, and rapidly obsolescent hardware and software. Increasingly, we are asked to do more with less. At the same time, the field of technical communication has grown so rapidly that it is virtually impossible to keep abreast of the many developments within and outside academe. For those who, like me, are in departments with only one or two faculty members specializing in technical communication, developing undergraduate and graduate programs in this field becomes exceptionally challenging, especially with the add-on responsibilities of supervising internship programs, running the student chapter of the Society for Technical Communication (STC), and helping place graduating students.

Let us assume, then, that no one is going to lighten our workload, that our hardware and software will always be a few versions behind industry standards, that the extra faculty members we need are not going to magically materialize, and that the field of technical communication will continue to evolve at dizzying speeds. Not an unrealistic scenario for most of us. Given this scenario, the question then becomes how do we improve the quality of our programs and give students meaningful educational experiences, without substantial additional resources and without putting further pressure on ourselves?

The key, as many of us have argued, is in exploring ways of involving others who most have a vested interest in the quality of students' education: those in government, business, and industry who eventually become students' employers. Paul Anderson (1995) has emphasized the importance of involving various stakeholders in our program planning, and Michael Keene (1997), in *Education in Scientific and Technical Communication: Academic Programs That Work,* has noted that the most successful programs feature internships, co-ops, practicums, and advisory boards. As for other types of affiliations, *Technical Communication* devoted an entire issue to the theme "Toward 2000: Education, the Society, and the Profession," and in one of the articles, Deborah S. Bosley (1995) described eight types of linkages between academia and industry and the benefits of each: classroom and curricula activities; student-faculty on-site opportunities; professional conferences and community organizations; employment and data research; corporate, textbook, and journal publications; liaison functions; and grants, funding, and donations. Most recently, the STC has also sponsored faculty internships in corporations as another way of making connections between academics and practitioners.

Certainly, the opportunities for collaboration have never been greater nor more important to the future of our profession for both practical and theoretic reasons.

INTEGRATING THEORY AND PRACTICE THROUGH COLLABORATION

For me, the motivation for helping develop this partnership had little to do with practical matters, although I am by nature a pragmatic person and although, as I alluded to previously, such partnerships can help departments solve problems related to insufficient staff, expertise, and budget in technical communication curricula. Rather, the partnership seemed an ideal way to do something that I have been trying to do in a number of ways over the decades—to meld theory and practice. Having been a professor for a decade and a consultant in business and industry for another decade and then a professor again for more than another decade (with some of those years overlapping), I have gained a good understanding of how both communities—the academic and the business communities—view the other and am therefore especially impassioned about the need to integrate these worlds. Much in these two cultures, however, militates against that integration. In fact, there is considerable dissension even within our academic community.

For example, in the last twenty years, since the Elizabeth Tebeaux (1980) and Elizabeth Harris (1980) exchanges in *College English,* the debates about whether theory or practice ought to drive our profession have occupied much of our time. In many ways, the debate today remains essentially the same, though framed in different terms: for example, whether technical writing is rhetorical or instrumental discourse (Miller 1996; Moore 1996) and whether students' first allegiance in the workplace ought to be to the organization's goals, through enculturation and assimilation (Tebeaux 1985), or to a sociopolitical agenda based on cultural criticism and change (Herndl 1993b; Johnson-Eilola 1996).

Added to the friction within our ranks is the gulf between those who teach in academe and those in the workplace (meaning, of course, the *other* workplace). A few years ago, George Hayhoe (1998) pointed out in *Technical Communication* that there seems to be a growing distrust between those who practice technical communication and those who teach and study it. Recently, Barbara Mirel and Rachel Spilka (2002) observed that industry relies very little on academic scholarship and that it has had "little or no impact on practice" (207). Essentially, the gap is, again, the old one between practice (knowing how) and theory (knowing what and why). At root, according to Stan Dicks (1999), the disconnect between academics and workplace professionals is due in part to the very different worlds each group inhabits. As he contended in his presentation at the1999 annual conference of the Association for Teachers of Technical Writing, academicians are not particularly suited by character, temperament, working conditions, or professional incentives to get along with professionals in the workplace, let alone with each other.

However, if we are going to help bridge the gap between theory and practice and between the workplace and the academy, we are going to have to learn—not just study, write, and talk about—collaboration and to do it well ourselves. And this learning means more than simply inviting guest speakers from the "real world" to address our classes, incorporating client projects into our classrooms and assignments, or setting up advisory boards, internships, or shadowing programs. Although all of these approaches are important planks in building that bridge (and most of us have been involved in such initiatives), something more is needed if we are going to truly understand each other and enhance our mutual professional interests.

I suggest that rather than waging turf wars with practicing professionals as to the primacy of theory or practice, we begin to engage in what Barbara Couture (1998) calls a phenomenological rhetoric, where we are open to others' experiences and interpretations; what Richard Young, Alan Becker, and Kenneth Pike (1970) call Rogerian rhetoric; and what James Berlin (1993) calls transactional rhetoric, where the goal is not persuasion but understanding, learning, cooperation, and professional growth. For, as Hayhoe (1998) warns us, "Without cross-fertilization, both academe and industry face the prospect of sterility" (20).

OVERVIEW OF THE NIU/CHICAGO CHAPTER STC INSTITUTE

For the last eight years, I have been involved in one such effort at cross-fertilization: the development of the Chicago Chapter STC Institute for Professional Development (hereafter, simply "Institute"), now governed by a board of those from academe, business, and industry. For the past six years, Northern Illinois University (NIU) has partnered with the Institute to offer courses planned and taught by researchers, teachers, and practitioners of technical communication from several universities and companies. The courses feature the following attributes:

- All-day Saturday course sessions, taught off campus near business and industry

- A mix of course enrollees, including traditional university students and workplace professionals
- Team teaching and review of participants' work by those in academe and business/industry
- Real-world client projects as assignments
- Mentors from business and industry assigned to work with course participants
- Undergraduate or graduate credit or continuing education units (CEUs)

How the Institute Began

The Institute was initiated eight years ago when a few members of the education committee of the Chicago chapter STC, of which I was a member, decided to try developing a partnership of practicing professionals and area university professors to offer courses for those new or relatively new to the field of technical communication. After two years of planning and an STC grant to provide seed money, the Institute was ready to pilot two courses: Fundamentals of Technical Communication, for those with no prior experience or course work in technical communication, and Topics in Technical Communication, for those with either some prior coursework or experience. Since then, these two courses have been offered annually through Northern Illinois University.

What the Institute Is

Officially, the Institute is an entity of the Chicago chapter STC and, as such, reports directly to the chapter's administrative council and, through them, to STC headquarters. The charter for the Institute was written by the members of the Chicago chapter's Institute committee and was approved by the Chicago chapter's administrative council in August 1998. The charter sets forth the Institute's mission, structure, and operating parameters; its relationship with groups and individuals outside the Institute and STC; and the Institute's governing board and five directorship positions.

The goal of the Institute is to provide a unique educational program for both entry- and experienced-level professionals by offering courses in technical communication that integrate academic theory and practical application. This combination of research-based principles and hands-on practice lies at the heart of the Institute and is its whole *raison d'être*. When the Institute was conceived, its founders were mindful of creating courses not already available through either educational institutions or vendors. If another group or school was already offering something similar or could do it better, then, we figured, there was no reason for us or the Institute to do it, too.

Affiliations

In support of this goal, the Institute may form relationships with individuals, academic institutions, consulting firms, and professional associations or organizations. From the beginning, the Institute was designed as a collaborative effort of teachers, researchers, and practicing professionals in the field of technical communication, and many individuals associated with the Institute have filled all three roles in their careers. In fact, it was just such a group of individuals who, while serving on the Chicago STC education committee, came up with the idea of the Institute in the first place. Now eight years later, the Institute carries on this tradition of collaboration at all levels—whether it is in the makeup of the board that governs the Institute; the instructors who design, plan, and team-teach the courses; or the course participants themselves, as they develop teamwork skills by working jointly on course projects and in-class application exercises.

Practitioners

Chicago has one of the largest STC chapters in the country and accordingly its members represent a wide range of businesses and organizations, big and small. The Institute invites those with the most experience and expertise to become instructors, mentors, panelists, or brown-bag speakers in its courses.

Underlying the Institute is the belief that there is no substitute for practical know-how. In the classroom, the practitioners bring the voice of experience to bear upon what might otherwise seem abstract theories, principles, and strategies. The practitioners can quickly draw upon a wealth of examples, stories, and case studies—all of which help establish the instructors' credibility and at the same time make the class more interesting and directly applicable to the participants' internships, client projects, and current or prospective jobs.

Academics

The Institute also encourages the involvement of those who teach, study, and research technical communication in colleges and universities. Thus far, individuals associated with DePaul University, Illinois Institute of Technology, Purdue University (Calumet), Illinois State University,

Northern Illinois University, and College of Lake County have served as instructors. Conceivably, the Institute board could extend its reach even farther and, perhaps someday, form a consortium among several schools.

In the practice/theory binary, if you think of the professional technical communicators upholding the practice end, then it is the academics—the professors, researchers, and scholars—who uphold the theory end. They are the ones whose job it is to study technical communication, to be familiar with the research, and to conduct the research—to find out not only what works and how it works (what the practitioners know best) but also why and in what situations it works, even to critically examine what we mean by "works"—a process that includes familiarity with the values underlying our actions and words, as well as their social and political implications.

Because the main goal of the Institute is to break down the practice/theory binary and remove the walls that separate the classroom and the workplace, the Institute screens its instructors carefully for those who understand the importance of informed practice on the one hand and grounded theory on the other. Thus, contrary to the old adage, "Those who can, do; and those who can't, teach," each of the instructors who teach for the Institute also has real-world experience, many as independent consultants. Others are former professors who have left academe to become full-time, corporate employees.

These, then, are the two main groups affiliated with the Institute. However, in terms of its mechanics and operations, there is another important player in this venture.

Partnership with Northern Illinois University

Since its beginning, the Institute has partnered with NIU. The reasons are many: some pragmatic, some economic, some ideological, and some personal, not the least of which is the fact that I am a professor at NIU and I also served on the original Chicago STC education committee from which the institute idea was conceived, on the subsequent Institute committee that developed it, and then as the first managing director of the Institute governing board.

Pragmatics. Most importantly, NIU gave the Institute courses an institutional home by providing me with some released time to serve as the course coordinator and by helping handle such administrative matters as course listings, registrations and records, credit, and course location. NIU's College of Liberal Arts and Sciences, the Department of English, and the Office of External Programming have been exceptionally supportive of the partnership.

Technically, the Institute courses are NIU courses because they are offered through the university and appear in the schedule of courses. Unlike other university courses, the Institute-affiliated courses are offered for either undergraduate or graduate credit or for CEUs. Furthermore, the courses can be taken for a grade or pass/fail or may be audited. This flexibility in credit and grade options makes the courses attractive to a wide audience, including traditional university students seeking degrees, nontraditional students fulfilling professional growth requirements for their jobs, and those both seeking a degree and working full time. However, because the courses involve active participation and rigorous reading and writing assignments, few have elected the audit or CEU options in the last six years. Half of the spaces in the Institute courses are reserved for non-NIU students.

The maximum enrollment in each course is twenty-five (with the expectation that a few will drop before the course begins), and typically the courses have had waiting lists. Most of the participants enroll for graduate credit, and the majority of these are students-at-large (that is., nondegree graduate students at the time of enrollment), some of whom plan to transfer the credit to other universities, some of whom later apply for admission to our graduate or certificate program in professional and technical communication, offered through the English department.

Economics. Another reason for the appeal of the university partnership is sheer economics. Compared with the cost of vendor courses or with the tuition at private universities in the area, NIU's tuition is quite reasonable—anywhere from a half to a third of these other costs. Also, NIU pays an honorarium to all of the instructors and mentors associated with the Institute courses, although not to the brown-bag speakers or panelists. The honoraria are insignificant when compared with the consulting fees that these professionals normally command and when compared with the costs and difficulty of adding additional, full-time faculty.

On the Chicago chapter STC side of the ledger are some shared marketing, printing, and postage costs. However, these costs have declined as the Institute has moved much of its publicity and orientation materials to the NIU and the Chicago chapter STC Web sites. All five directors of the Institute board are volunteers from industry and academe.

How the Courses Are Structured

For six years, the Institute has offered two courses through NIU: Fundamentals of Technical Communication (FTC) in the spring and Topics in Technical Communication (TTC) in the fall. Although neither is a prerequisite for the other, the preferred sequencing would be to take the FTC course first and the TTC course second. If you are wondering why the more basic course would be offered in the second semester (spring) and not the first semester when the academic year begins, the answer is simple: to allow sufficient time in the fall to promote the spring FTC course, especially to market it to the current and prospective members of the Chicago STC. Summer is not a good time to advertise the course because most students and faculty are gone and because the Chicago STC doesn't hold chapter meetings or have newsletters in the summer.

Neither course is a tools course because those types of workshops (for example, RoboHELP, PageMaker, and Dreamweaver) are readily available in the Chicago area and, moreover, fall outside the Institute's purview. Since the Institute started, the courses have met in a "smart classroom," with an instructor podium well equipped with a computer, projection system, TVs, VCRs, and a direct connection to the Internet. To project their materials, most instructors either plug their laptops into the podium or bring Zip discs or CDs. In the past couple of years, especially now that NIU has a new suburban campus outfitted with the latest technology, many classes incorporate computer-lab sessions, interspersed as part of the small- group activities.

Typically, the class sessions involve a combination of lecture, discussion, and application exercises. In our orientation sessions with the instructional team and in our instructor guide, we stress the importance of student participation and interaction. The layout of the classrooms and computer labs facilitates this importance, with comfortable chairs that can be easily rolled to different tables or workstations. Also, the building in which the class is held has several other rooms and lounges to accommodate breakout groups.

For both courses, the Institute board prepares orientation materials and holds orientation sessions for new instructors. In the FTC course, there are also orientation materials and sessions for the mentors. In both courses, the Institute board works with the instructors to develop customized course materials. Because this course is STC affiliated, we rely heavily upon articles from *Technical Communication* and *Intercom.* In each course, usually one or two textbooks are also required or recommended.

Because there is not space here to describe both courses in detail, I will spend more time on the first one, the FTC course, and will only gloss over the main features of the second one, the TTC course.

Fundamentals of Technical Communication (FTC) Course

The Fundamentals of Technical Communication (FTC) course is an introductory course designed for those new to the profession. It is offered for seven Saturdays from 9 A.M. to 4 P.M. in a suburb of Chicago. The course has a maximum enrollment of twenty-five participants and carries either three semester-hour credits (at the undergraduate or graduate level) or four CEUs. It features a team of instructors, some from academe and some from business and industry, each team member responsible for a different topic and for teaching one of the Saturdays. In addition, a university faculty member (me in all FTC courses so far) coordinates the entire course, attends all sessions, and works with each course participant, along with that person's assigned mentor from business and industry. There are four mentors total, who work with five or six protégés each.

Course Topics. The FTC topics include an introduction to the profession and discipline of technical communication; basic skills, knowledge, and attitudes necessary for success; writing clearly, concisely, and correctly; organizing and presenting information; usability testing; document design for paper and electronic media; and ethics and technical communication. Instructors of these sessions, who also receive advance copies of the participants' project proposals, are encouraged to make connections between their topics and the participants' client projects. During the last two Saturday mornings of the course, the participants give short oral presentations describing their projects, share what they learned from doing them, and suggest what they might do differently were they to do their projects now, knowing what they did not know before.

Scheduling. The FTC course begins in mid-March, when the official spring semester at NIU is half over. Although the FTC course skips the first half of the semester, by meeting all day for seven Saturdays we still log at least the same number of hours as a regular-semester NIU course: in our case, probably more than fifty total contact hours. These hours include individual conferences before and after class and some brown-bag lunch sessions on topics requested by the course participants. The lunchtime speakers are other technical communicators and STC members who want to become involved in the Institute but with a minimal time commitment at this point.

Preliminary Contacts and Client Projects. As the course professor/coordinator, I correspond with all of the enrollees about a month before the course begins to explain the main requirements and to give them guidelines for finding a real-world client project to work on throughout the course, for conducting a client interview, and for writing a project proposal.

The types of projects span the spectrum of technical writing: from brochures and newsletters to manuals, formal reports, online documentation, and Web pages. Clients include present employers, friends, neighbors, clubs, churches, small companies, and large corporations—virtually anyone who has a genuine need to have something written. Projects developed for nonprofit organizations are especially encouraged because they carry low risks but high rewards—namely, the satisfaction of having volunteered one's services for a worthy cause. The participants are not paid for work done in conjunction with their course projects.

By the first session, each participant has submitted a project proposal and participant survey to me and the course mentors and instructors. The directors of our Institute board use these to match each course participant with one of the four mentors whom we have arranged to have work with them over the next seven weeks on both their client project and their professional development goals and plans. In addition to working with a mentor, each participant consults with me at various stages of the client project.

Electronic Journals. Another way in which the mentors and I stay in tune with the participants is through their electronic journals. Soon after each session, each participant sends the mentor and me reflections about the class and the preparatory readings. In these e-journal entries, which are intended to be spontaneous and candid, the participants also relate the specific day's topic to their client projects and career goals. Participants are encouraged to develop a personal inventory of the skills they think they need to develop to prepare themselves as technical communicators, as well as a plan for their professional growth. These e-journals often become the springboard for topics of discussion during individual conferences with the instructors, mentors, or me; during panel presentations in class or brown-bag lunch sessions; or during informal chats over breaks and lunch. The bonding that occurs throughout the seven weeks is exceptional, to the extent that many participants continue to see each other and have even hired or contracted each other. Many have even asked the Institute to hold reunions, the first of which was held a couple of years ago during an FTC first-Saturday luncheon, followed by a panel of the FTC alums, now themselves employers, thereby completing the Institute circle.

Research-Related Projects. Finally, for those enrolled for graduate credit, another requirement is a research-related project, which may be done in conjunction with the main client project or may involve inquiry into another topic of importance to them and to the field of technical communication. Some have conducted primary research (mainly interviews or surveys), and others have done traditional, secondary research. We always encourage the participants to make their deliverables something of use to someone else (such as a report, feasibility study, or proposal for a company or organization; a book review or article for a journal), instead of the usual research paper that simply gets filed or tossed out once it has been turned in. We thereby reinforce the principal theory upon which the Institute is based: that writing is not an end product but an integral part of a process involving human interaction and that, accordingly, writing shapes as well as reflects our beliefs, behavior, attitudes, and institutions.

Evaluations. As NIU's instructor of record and as the course coordinator, I evaluate each project and am responsible for submitting final grades. In my assessment, however, I consult with the mentors to enlist their input.

As for evaluating each session and instructor, participants complete an evaluation form at the end of each session, copies of which are given to the respective instructor, as well as to the Institute board (and, thus, the Chicago chapter STC administrative council) and NIU. At the end of the course, participants also evaluate both the course as a whole and their mentors. And, finally, mentors and instructors evaluate their roles in the course and offer suggestions for improvement. The Institute board reviews each of these evaluations in its course planning for the following year.

Topics in Technical Communication (TTC) Course

The second course, offered each fall, is Topics in Technical Communication (TTC), which is designed for those with either some prior coursework in technical communication or some prior experience (one to three years). For those who may be interested in developing a similar course, perhaps apart from a two-course sequence like the FTC/TTC Institute offerings, I describe both how the TTC course used to be structured and how it is now structured.

Previous TTC Format. For the first couple of years, TTC was structured in much the same way as FTC: that is, for seven Saturdays and three credits (again, undergraduate or graduate credit or CEUs). The topics and instructors were different from those of the FTC course and usually changed each year. For example, during the first year, topics included Analyzing Business Needs and Developing Solutions; Planning and Managing Projects; Thinking Visually; Developing Online Documentation; Presenting Complex Information Verbally; and Employing Research Strategies for Corporate Projects. Subsequent years' topics included Design, Documentation, and the Product Life Cycle; Moving Documentation from Paper to Online; Information Modeling; Technical Editing; Web Design from the User Perspective; Interviewing Subject Matter Experts; Designing Newsletters; and Designing Computer-Based Training. Some topics spanned one and a half or two Saturdays, with the same instructor returning to class to review the work submitted during the interval. Other topics that did not have associated homework were covered in one-day or half-day sessions.

Current TTC Format. For the past three years, the TTC course has shifted to a modular format, offering four different minicourses, each on a different topic. Each module or minicourse lasts two Saturdays from 9 A.M. to 4 P.M. with at least one Saturday in between to permit time for a written assignment based on the prior Saturday's lesson. Each minicourse is offered through NIU and carries one semester-hour credit at the undergraduate or graduate level or the equivalent CEUs. (No one to date, however, has chosen CEUs over university credits.) Like the FTC course, the TTC course is held at NIU's Chicago suburban location (in one of the main high-tech corporate corridors) and draws both traditional and nontraditional students (mainly working professionals). The enrollment maximum for each minicourse is twenty.

A different instructor plans and teaches each minicourse, and the minicourses are independent from each other. Thus, individuals can enroll in one, two, three, or all four courses for an equal number of semester-hour credits, which can be applied to NIU's or, when permitted, to another institution's degree or certificate programs. As with the FTC course, the instructors are selected by the Institute board and hired by NIU as guest lecturers, who are paid an honorarium by our Office of External Programming. At NIU, this arrangement avoids having to maneuver through the administrative maze necessitated by hiring an adjunct or visiting professor (a process that can take up to two years). Thus, an NIU English department professor serves as the instructor of record for the minicourses. This person, who also happens to be the Institute's managing director (formerly me, now Philip Eubanks), coordinates the planning of the minicourses, along with the Institute board; corresponds with the instructors; reviews all of the written work submitted by the minicourse participants, which the minicourse instructor has assigned, evaluated, and graded; and completes the final grade sheets for all minicourses.

Finally, it is important to note that this change in the TTC format (from a required set of topics and credit hours over seven Saturdays to modular topics from which individuals could choose) was precipitated by surveys and focus groups involving the course participants themselves. Furthermore, before changing the format entirely, the Institute board planned and piloted only one such minicourse during a fall semester when the original TTC course format (seven Saturdays) was still in place. Because the pilot was well received, the Institute board decided to forgo the original format to permit greater flexibility of scheduling and variety of topics. The Institute continually tries to practice what it preaches in the courses: that the first principle of good communication is to meet the audience's needs.

BENEFITS OF THE PARTNERSHIP

As I hope is evident from this description, the partnership between the Chicago Chapter STC Institute for Professional Development and Northern Illinois University is meeting an educational need that has long been recognized by both academics and practitioners of technical communication: the need to integrate theory and practice. Because the Institute courses are planned and taught by a team of professors and practitioners and because the courses are incorporated into the undergraduate and graduate curricula of the university, they are intellectually rigorous and pedagogically rich. In fact, some of the other colleges and universities whose professors are affiliated with the Institute are now examining the possibility of accepting the Institute courses for transfer credit into their own institutions. As a result, in the future the Institute may well develop into a consortium of many colleges, universities, businesses, industries, and not-for-profit organizations.

The private sector's matrix model of project management and cross-functional teams—of sharing resources, talent, and expertise—is already influencing education, and we have much to learn from it. Colleges and universities can no longer afford to both create and duplicate programs in technical communication. And even if we could, we would never be able to keep pace with the advancements on the professional-practitioner front. For some, the answer to educating future technical communicators is to develop distance-learning courses. Although I would be the first to admit that distance learning has its place in helping to provide educational access to great numbers of students in far-off locations, it cannot substitute for real-time, face-to- face, human interaction, collaboration, and situated learning such as that afforded by the type of partnership described here.

However, for my colleagues at institutions situated in more remote parts of this country, or other countries, I can easily fathom migrating the concept of the NIU/Chicago Chapter STC Institute for Professional Development to the Internet. Imagine it—a virtual consortium of top-notch professors, teachers, scholars, and practitioners collectively planning, teaching, conferencing, and mentoring in a cybercollaboratorium.

SOME RECOMMENDATIONS

For those in academe seeking to establish similar types of partnerships with professional organizations, I could write an entire book based on my experiences over the past eight years because there are so many things to consider and to avoid. But for those just starting out, I have a few important suggestions.

Network

Make connections with professional organizations to garner support for developing academe-industry alliances. Volunteer to become a member of their education committees. By affiliating with a professional organization or several of them, you have access to many practitioners who, once they know of your genuine interest in their expertise, are likely to welcome the opportunity to collaborate with colleges and universities. These individuals may well become the instructors, mentors, and guest speakers in any cosponsored courses you develop in the future. Furthermore, with the support of a professional organization, you are less likely to encounter the politics, special interests, and indebtedness that usually accompany partnerships formed with individual corporations.

Find Stakeholders

At the same time you network, get buy-in from your home institution. Although it is sound practice to get the blessing of your department chair and colleagues, oftentimes support may be more readily forthcoming at the higher levels of an institution—from, for example, your dean, president, or provost. These individuals are often more receptive to academe-industry affiliations because such ventures help give the institution visibility, credibility, and currency among its stakeholders at the city, county, and state levels. At a time when colleges and universities are being held accountable for how well they help prepare the workforce, such programs and partnerships are likely to get a good hearing. Furthermore, such academe-industry collaborations are looked upon favorably by alumni and granting agencies and thus can become the stepping-stone to additional revenue.

Explore Existing Resources

Existing resources may be internal or external. Before launching into something grandiose, assess what you're getting into and who may be able to help in the enterprise. Internally, find out what other types of partnerships your institution is engaged in and the extent of the support being offered them—whether, for example, it's a matter of personnel, marketing, technology, or printing costs. Externally, find out what other institutions and professional organizations are doing or have done. It may well save you a lot of time and frustration, and it may help provide you with other ideas and models.

Analyze the Partnership's Audience and Purpose

Although this analysis should be a no-brainer for us, it's surprising how often the obvious gets overlooked. This is not the time for a "build-it-and-they-will-come" approach. Survey the target audience to determine the level of interest and preferences for such things as course topics, structure, location, days, and times to ensure that what you're creating reflects genuine needs. Then reassess your assumptions periodically. Also, clarify what the respective partners stand to gain and lose in the arrangement. And, of course, commit as much as possible to writing (document, document, document) to clarify roles and responsibilities and to avoid confusion and misunderstanding later on. The NIU/Chicago Chapter STC Institute began with mail and telephone surveys of the Chicago chapter STC membership.

Set Priorities

Realize that setting up a partnership such as the one I described takes a considerable amount of time and effort, which will have to come from somewhere—unfortunately, usually from one's research unless it is possible to turn the partnership project into a research-related venture. Multilevel collaborating can be time consuming. The greater the number of people involved, the richer, and potentially more frustrating, the enterprise. Such projects have a way of either spiraling out of control or losing momentum unless carefully planned in stages. In the beginning, it's far better to develop a pilot course and test, than to create a full-blown program. KISS (Keep It Simple and Succinct) is still the best advice.

The NIU/Chicago Chapter STC Institute for Professional Development is not a one-size-fits- all partnership, but I hope that it will inspire those affiliated with other technical communication programs to consider the benefits of thinking creatively about ways to involve practicing professionals and other academics in planning and implementing collaborative courses. Partnerships are one way that professional organizations can help promote education and the profession; that colleges and universities can conserve resources by tapping into a rich reservoir of talent and expertise; and that faculty, students, and practitioners alike can expand their knowledge, skills, and opportunities. It can be win/win situation for everyone.

ACKNOWLEDGMENTS

The members of the Chicago chapter STC education committee (and subsequent orientation committee) that planned the Institute were Christine Abbott, Mary Ryba Knepper, Alex Kantis, Gina Meyers, and

Anna Miller. The directors of the current Institute governing board are Philip Eubanks, Managing Director; Brian Everett, Faculty Director; Roger Graves, Curriculum Director; Jaysa Jackson, Marketing Director; and Jennifer Kroll, Administrative Director.

16

MAKING CONNECTIONS IN SECONDARY EDUCATION

Document Exchange between Technical Writing Classes and High School English Classes

Annmarie Guzy
Laura A. Sullivan

CHAPTER FOCUS

Audience analysis is an important component of any writing course, but it is crucial to technical writing instruction. As defined by Lisa Ede (1984) in "Audience: An Introduction to Research," audience analysis involves "methods designed to enable speakers and writers to draw inferences about the experiences, beliefs, and attitudes of an audience" (140). Through the efforts of composition programs and the growth of the writing-across-the-curriculum movement, students have many opportunities to develop their writing skills in a variety of contexts. The majority of these tasks are academic in nature, and audience analysis, if performed, consists of assessing what knowledge the instructor wants students to demonstrate through the writing task, such as research papers and essay exam responses, and of making accommodations for that instructor's particular stylistic or format requirements. In contrast, typical assignments in technical writing courses—correspondence, instructions, proposals, and so forth—require students to envision specific audiences outside the classroom for whom to write and to consider the contexts in which these documents will be read.

Making the transition from academic to nonacademic audiences can be difficult for technical writing students, especially when the immediate document cycle is essentially the same as for other academic writing tasks: the student completes a document and submits it to the instructor to be evaluated for a grade. To increase focus on audience analysis, some features of nonacademic writing can be simulated in the classroom to give students a more authentic feel for nonacademic writing genres and processes; for example, peer critique groups can serve as editing panels or user-testing groups, and instructors can exchange documents between classes or with each other's classes to provide a separate group of readers. However, more instructors are now designing assignments in which students, either individually or in groups, must work outside the classroom with an on-campus or off-campus organization to create documentation. Some instructors provide guidance for these projects by soliciting organizations themselves, while others leave this responsibility to the students; in either case, nonprofit organizations tend to be productive choices because *(a)* assistance from students is welcome community service and *(b)* students are less likely to become involved in corporate politics during what is in essence still a relatively short-term project being completed as a component of an academic course. In any case, by taking writing out of the classroom and bringing it into the community, students gain authentic audiences, experience document cycles firsthand, and learn the importance of community responsibility.

However, some instructors and administrators are beginning to express concerns regarding safety and liability when requiring students to travel off campus for class assignments. Students are usually required to complete liability forms for off-campus field trips, conference trips, and so forth, but what are students' rights and schools' responsibilities regarding required off-campus study? Do students have the right to refuse to travel off campus to complete an assignment if they feel uncomfortable with their organizational representatives or their peer group, or is this position considered another component of adjusting to writing in the "real world"?

To address these concerns while still providing a service for an off-campus audience, we have organized document exchanges between Guzy's university-level technical writing classes and Sullivan's high school senior English classes. We began our exchange program after discussing the formal introduction of audience analysis into the curriculum in Sullivan's department and comparing that with the importance of audience in Guzy's technical writing assignments. Exchange cycles are relatively brief with students in one class sending documents either consisting solely of or including correspondence and the students in the other class responding to those documents; logistically, we have kept these brief so as not to interfere with other assignments required by the respective curricula.

We feel that our exchange is an innovative approach to teaching technical and professional writing, in several ways. First, this approach is innovative to the high school teacher because it gives the students a real audience, not a "teacher" audience that lacks the realism of an authentic audience. In this light, high school students stand to be greatly influenced in that their work will be authenticated; college students will benefit from this authentication as well, but will also gain experience in targeting their technical knowledge for a more general knowledge base. Thus, when performing audience analysis, students at each end of the exchange must consider the age and relevant academic experience of their exchange partners; to simulate a more global workplace, they also learn about another community in a different geographical and socioeconomic part of the country. Correspondence between exchange partners usually includes discussion of academic experiences and professional concerns, so students have the added opportunity for self-reflection and assistance in decision making as graduation nears for both groups. In doing so, students still write for an off-campus audience but from the relative safety of classrooms. Finally, the documents produced for the exchange require extreme diligence because, once sent, they are truly gone without students being able to intercept them to make corrections. This process is important for students because they rely so much on the document cycle of teachers receiving papers, correcting them, and handing them back for revision after revision.

PEDAGOGICAL/THEORETICAL FRAMEWORK

Research on audience analysis is plentiful in rhetoric and composition and has readily been adapted to theory and pedagogy in technical communication. For example, Coney (1997) reviews work on audience by a variety of groups, such as reader response theorists, cognitive psychologists, and ethnographers and concludes that "if any generalization is possible at this point, it is that for technical communicators, nothing matters *more* than audience" (5). For our exchange, we are working with interpretations and applications of this research at both the secondary and postsecondary education levels.

One of the foundations of contemporary research in audience analysis is the germinal work of Ede and Lunsford (1984), and, for our exchanges, their distinction between "audience addressed" and "audience invoked" is particularly important. Introducing these terms, the authors note that "[t]he 'addressed' audience refers to those actual or real-life people who read a discourse, while the 'invoked' audience refers to the audience called up or imagined by the writer" (156). For most technical writing assignments, students are asked to address a specific audience, such as the users for a technical manual, but in most cases, they are never actually able to communicate directly with their intended readers, to solicit feedback for improving the design and usability of the document. In fact, unless the student is actively engaged in a writing assignment for an organization outside the classroom, the only audience truly "addressed," according to Ede and Lunsford's definition, is the instructor who reads and evaluates the assignment. Classmates are also directly addressed when students give oral presentations, but through peer critiquing they may be more secondary readers and editors than primary readers for assignment purposes. With the secondary setting, students call upon the invoked audience for most writing assignments. Then, after completion and grade assignment, students file the papers away without taking value of the critique because there was not a real audience; therefore it is not of real concern to them. The students translate a "fake" audience to a "fake" experience, thus not giving credence to the assignment and, in effect, wasting paper and time.

Addressing this conflict of audience and purpose further within technical communication, Redish (1997) applies basic concepts from cognitive psychology to technical communication pedagogy, identifying four factors that aid in audience analysis and interpretation of technical texts:

> Many readers share experiences and, therefore, have similar schemas.
> The text (or product) influences and constrains readers' interpretations.
> Guidelines derived from empirical research can help technical communicators meet their readers' needs.
> Techniques exist for getting feedback from audiences on draft materials. (73–74)

One of the most important problems technical writing students have with audience analysis is identified in Redish's first factor: students' schemas concerning writing are based in academic writing tasks for their instructors, and attempting to balance writing for the technical writing instructor with writing for an "addressed" audience can cause cognitive dissonance: "Students may develop schemas that tell them there is one truth in the classroom and another in the real world" (78). The secondary students regard the classroom experience as far removed from real world applications as well. Instructors need to help students adapt the schema of a "false" setting within the classroom to a "real-life" situation within the workplace environments. This situation can help students to bridge the gap between academia and the workplace, thus adding credibility to classroom instruction. Once we have established this reality, students can readily adapt the schema to match workplace goals, which, in turn, makes them more equipped for the ever changing workplace. Proposing ways in which instructors can help students change their interpretive schemas concerning technical communication, Redish charges instructors to actively engage students in learning while raising an important question about the conflicting schema:

> Lecturing at students seldom results in real learning. But activity by itself is also not enough. The activity has to be situated in realistic contexts. . . . [But a]re the students getting mixed messages? Are they being told that they should write for a 'real audience' when they know that their product will be read only by an instructor who is not part of the real audience? (80)

By creating our document exchange, we have given students the opportunity to write to a "real," actively addressed audience. We still review the documents as instructors to monitor general content and quality of the material being exchanged (see the "Implications" section for potential conflicts with grading and openness between students in exchanges), but we are no longer the primary audience for the documents. We work from class rosters to match academic and professional interests as closely as possible, so students have names of people to whom they can specifically address correspondence.

Having an individual to whom to write begins the audience-analysis process, but it continues in a unique way when students must acknowledge that they are writing for people at different levels of academic development. As Ede and Lunsford (1984) state,

> Even the conscious decision to accede to the expectations of a particular addressed audience may not always be carried out; unconscious psychological resistance, incomplete understanding, or inadequately developed ability may prevent the writer from following through with the decision—a reality confirmed by composition teachers with each new set of essays. (166–167)

The problems Ede and Lunsford identify are typical of problems students have during the exchange. First, psychological resistance, and rather conscious resistance at that, may come in the form of students not taking the assignment seriously because *(a)* the instructors are not the primary audience, *(b)* the assignment seems so different from the rest of the coursework, *(c)* the audience seems contrived and not one they would write to in their future school or work settings, and, *(d)* quite mundanely, the exchange may not figure prominently in their final grades. Therefore, we encourage them to include relevant personal information such as discussion of extracurricular activities and hobbies, as well as questions about what to do after graduation and what career paths to pursue. Incorporating such information can motivate students to be more active writers and thoughtful readers and responders; Allen (1989) argues that

> motivation may make significant differences in the way readers read: the amount of time they are willing to invest in reading a document, their attitude toward it, the kinds of information they are most likely to garner from it, their expectations of it, and other factors that would require the technical writer to go far beyond the concerns of traditional audience analysis. (54)

Incomplete understanding and inadequately developed ability, however, are more pressing problems that may be identified, although not fully solved, only during the course of the exchange process. Students at both ends may have problems with their writing, particularly with grammar and mechanics; and adapting their writing style and content for their respective audiences can be challenging when crossing educational levels. This challenge is especially true for some college juniors and seniors who have naturally become heavily involved in their major coursework but who may not be easily able to explain what they do without using heavy doses of technical jargon and abstract concepts. As Allen attests, audience analysis depends on "the technical writer's understanding of what the audience can be expected to know, enabling the writer to determine where to add detail, descriptions, definitions, analogies, or other aids to understanding" (58).

To address this problem at the college level, we follow Redish's third suggestion for using empirical research to help writers meet readers' needs. Burnett's (1997) textbook *Technical Communication* includes a chapter on audience in which she identifies and describes several types of prospective audiences for technical writing and useful sections for younger audiences. The first section we discuss is Burnett's description of students:

Students, from those in advanced high school courses to those majoring in technical subjects in college, have a particular interest in technical material. They read as part of their academic preparation or their cooperative work-study programs. They are interested in learning disciplinary knowledge and forming opinions to gain a broad background and eventually to become professionals in a specialized field. Subjects range from metal optics and metal fabricating to biomedical research and oceanography.
Familiarity: often know generalizations in a field; typically need information that provides technical details as well as implications
Expectations: information that will help them with *assessing, learning, learning to do,* and *doing;* usually interested in theory as well as practice
Typical education: may have already completed high school or undergraduate programs; may have specialized training from summer or part-time jobs, internships, or co-op programs. (68)

However, for some of the college-level technical writers, this discussion does not emphasize enough the importance of "translating" information about their majors and research interests to their correspondents; therefore, we also review Burnett's description of children:

Children are increasingly reading technical documents adapted to their level. Many of them enthusiastically read science books and have their own subscriptions to children's science magazines. They also read the technical documents that come with their computers, models, and video games.
Familiarity: may know generalizations in a field; want information that explains how and why things happen; need special consideration for limited formal concepts and vocabulary
Expectations: information that will help them with *learning, learning to do,* and *doing;* widely varied interests, sometimes wanting background information and at other times wanting help completing a task
Typical education: may have completed elementary, middle, or junior high school; may have specialized knowledge from hobbies and activities. (69)

In addition to discussing the importance of editing jargon and defining field-specific knowledge in more detail, we also expanded our audience analysis beyond age and educational level to include geographic and socioeconomic differences that might influence the students' writing and their readers' perceptions. Allen (1989) argues that technical writers need to consider effects of geographic changes and social and cultural differences (62); with advances in electronic communication and the global economy, students need to enter their future workplaces ready to acknowledge and address differences in communication styles of people outside their local discourse communities. As the audience analysis allows students to learn about different areas of the country—Midwest and Southwest, industrial and agrarian, rural and suburban—students also have the added benefit of observing the similarities between their immediate communities. In an intrinsic yet reassuring way, discovering similarities in academic and professional interest concerns with someone halfway across the country not only demonstrates to students that they are not alone in their postgraduation concerns but also opens students' minds to academic and professional opportunities outside their hometowns and states. Students at the secondary level typically are closed to the reality of what lies ahead of them in the real world and often hide behind the familiar walls of the high school. To help transition them to the real world, this exchange shows them a life beyond the security of home and helps to condition them to a larger world waiting for them. By giving them a taste of college and what is expected there, the exchange helps to cushion the blow by giving them a safe taste without the fear of failure. Also, for many of these students, life beyond their city limits may seem impossible and overwhelming, but the exchange opens the world to them, allowing a greater sense of reality to begin to seep into their consciousness.

DESCRIPTION AND ELABORATION

On one end of the document exchange is Sullivan's annual spring senior composition class at Granite City High School in Granite City, Illinois. Granite City is a suburb of St. Louis, Missouri, with a population of approximately thirty-five thousand. Its residents' economic status tends to be working class to middle class with one prominent employer being Granite City Steel. The high school population is approximately 2,200 students.

On the other end of our first two exchanges was Guzy's annual spring advanced technical and professional communication class at New Mexico State University (NMSU), located in Las Cruces, New Mexico, about forty-five minutes from El Paso, Texas, and the United States–Mexico border. Las Cruces has an approximate population of seventy thousand residents, and the town's economic status shows a divide between more affluent employees of the university and local aerospace and military employers (NASA, White Sands Missile Range) and the area's farmers, ranchers, and migrant workers. In fact, in 1994, Doña Ana county was the fourth poorest county in the United States. Many of NMSU's fifteen thousand students come from Las Cruces and from smaller rural towns throughout New Mexico. The advanced technical and professional communication course is required for juniors and seniors in NMSU's Wildlife Science and Chemical Engineering programs, so these majors are represented prominently in the exchanges; other typical majors enrolled include premed, nursing, and other natural and applied science fields.

To begin each exchange, Sullivan and Guzy assisted students with audience analysis and preparation of documents through classroom exercises (see appendix for exchange schedule). First, reviewing chamber of commerce–style materials about their respective cities and schools led to class discussions about cultural and socioeconomic differences between the areas and the stereotypes students might have held about their audience. For example, NMSU students thought that people "back East" did not know that New Mexico was part of the United States (although sometimes this is actually true). Then, the students studied the conventions of professional correspondence to write letters in which they discussed academic, professional, and personal interests. In some cases, students attached additional documents for their correspondents' review, and these documents will be addressed in the respective exchange narratives discussed later. To prepare students for the exchange, Sullivan led her students in a discussion about postsecondary options available to them. This exchange created an atmosphere of acceptance of the papers, much like introducing a new baby home to an older sibling and made students aware that the papers arriving would not be polished papers from teachers and professionals, but from peers with a little more education. This understanding helped to establish a bond early on, so students were more receptive to the idea of the exchange.

Spring 1997

The first document exchange began at NMSU. In one thread of assignments, Guzy required students to select topics from their majors and take these through several genres, beginning with a three-page position paper and annotated bibliography on an issue and ending with a lengthier proposal and oral presentation. For the exchange, Guzy requested that students revise their position papers for their new readers; the students' main challenge was to perform enough revision of technical concepts and terms that the high school readers could understand the content of the documents. For example, one student writing about New Mexico's Waste Isolation Pilot Project (WIPP) added description to help define the problem he was addressing:

> *Original Position Paper material:* At the present time in the United States of America there does not exist a central repository for the long term storage of high level nuclear waste. Most of the high level nuclear waste that is generated by both the civilian and military sectors is stored on site, in temporary storage facilities. This condition has lasted for more than fifty years now, with most of the temporary storage sites now filled to capacity and beyond. High level nuclear waste can be separated into two different types: high level waste and transuranic waste. High level waste consists of the spent fuel rods from civilian and military reactors, reprocessed fuel rods and radioactive material recovered from obsolete nuclear weapons. Transuranic waste includes compounds such as plutonium, by-products from nuclear weapons production and certain laboratory grade materials.
>
> *Revision addition:* The problem of nuclear waste disposal in this country can be compared to local trash collection. For instance, consider the trash collection in your own hometown. Every week the city comes and picks up your trash and takes it to a landfill or other facility for proper disposal. Now consider what would happen if the city did not come and pick up this trash. You would be forced to dispose of it yourself or store it in your house. Eventually, you would run out of living space and the threat of disease would be such that you could no longer live in that location anymore. This scenario is similar to what is happening with the nuclear waste in this country.

Guzy's class had also covered correspondence and resumes in an earlier unit, and Sullivan's class was going to begin that unit shortly, so Guzy's students included their resumes and wrote brief cover letters, which discussed not only their position paper topics but also the academic and professional choices reflected on their resumes. For instance, the student previously mentioned included the following text in his letter:

> The following report is a discussion of the problem of nuclear waste disposal in this country. This sounds like a very difficult and boring subject, but if one

> looks at the information with an open mind, the subject is not that difficult or boring. I hope that you find this report useful and informative.
>
> A little background on the author of this report. I am a senior studying Chemical Engineering at New Mexico State University. I will be graduating this semester and am looking forward to starting my career. I am thirty years old, yes that's right, I am an old man. Before coming back to school, I spent ten years in the United States Navy. It was in the Navy I learned all about Nuclear waste. I am in fact still in the Navy, as a Lieutenant Commander in the reserves. Upon graduation, I am going to be going to work for the Intel Corporation as a Process Engineer working on developing new microprocessors for personal computers.

All of the documents were then mailed to Granite City High School. Sullivan's students were then able to use the resumes and letters as design templates for assignments. For example, the student assigned to read the WIPP paper wrote this letter:

> I am writing in response to your letter and report. Although nuclear waste disposal is not my field of expertise, I found your report to be rather informational and interesting. I greatly appreciate the time and effort that was put into your project.
>
> A little about me, as a person and student. I am eighteen years of age and am currently a senior at Granite City Senior High School. I write for the school paper and have participated in several school sports. I will be graduating this June and look forward to attending Illinois College. I currently have a B average and hope to carry that with me through college. I plan to study medicine at IC and will, with the permission of grades and money, attend Med. School shortly after graduating. The choice of field is uncertain at this point however, but I feel that I have several years to choose one.
>
> I would like to wish you luck with your future as a Process Engineer and congratulate you on a report well written!

Several of the position papers, however, proved more troublesome because some of the technical writing students had given less attention to revising their documents than others had. For example, this position paper material was not revised at all from the original document:

> By using recycled PET [polyethylene terephthalate], not only does it reduce production cost compared to newly manufactured PET, but the growing environmental problem of long term disposal of plastic waste is also reduced. In

> order for the PET to be usable it has to be broken down into a form in which the cement and it can mix. The plastic must be transformed into a liquid but still maintain the basic original chemical structure. The plastic contains long rigid chains of molecules that make up the solid. The composition has to be altered in order to make a liquid; such a process is called depolymerization. With PET this is done by combining the PET with glycols of either ethyl propylene or neopental. A glycol is a chemical compound that contains two oxygen-hydrogen groups on the molecule. With sufficient heating the long chains of the polymer are broken down into their basic monomer structures. A monomer is the ionically stable building block of a polymer. The monomers are then reacted with anhydrides to form unsaturated polyester resins. Unsaturated polyester resins are molecules that have carbon-carbon double bonds. This somewhat deactivates the polymerizing properties of the monomer. The depolymerized resin is usually very viscous and needs to be diluted with styrene. By diluting, the resin can be mixed easier and further cured to give a harder finish.

A few of Sullivan's students were able to use the information from the position papers for senior term papers, but overall, the response rate was fairly low: less than half of the technical writing students received handwritten reply letters.

In our review of this first exchange, we felt that the results were less successful than we had hoped. First, we had discussed the exchange before the beginning of our respective terms, but we did not decide upon what to include in the exchange in enough time to include it officially in our respective curricula. Therefore, students ended up working on exchange materials in addition to the regular course assignments, which may have contributed to a lack of focus in these materials. Also, without enough substantial revision, the position papers were relatively confusing and intimidating to the high school students. As we had begun the exchange relatively late in the term, right after March midterms, the time needed on both ends for writing, revising, and mailing through regular U.S. postal mail at such relatively short notice for the students affected the quality of material on both ends.

Spring 1998

For the second exchange, we began in early February, shortly after the high school's new semester had begun. Sullivan now had personal email, so she requested and received a copy of Guzy's class roster with respective names, majors, and grade levels. This exchange then began with Sullivan's students writing typed or word-processed letters personally addressed to Guzy's students. Letters included descriptions of academic and social activities and questions about jobs, school, and local culture. For example:

> Hello, my name is ——. I live in Granite City, IL and I am currently a senior at Granite City High School. I play varsity for the high school soccer team and I also play select soccer.
>
> I plan on attending Lewis and Clark Community College then moving on somewhere better. My grades aren't very good, but I'm a good athlete. Hopefully I'll do better than I would in a bigger school. I'm kind of nervous about going to college and starting out at a close place would help.
>
> The best thing about college is going to be that I'll get to move out. My parents will probably kick me out. My dad is a lawyer and my mom is a nurse. I don't have any brothers or sisters but I have a dog.
>
> Well, that's about all I can say about myself, I hope you'll write me back and tell me about yourself.

The exchange was also now officially on Guzy's technical writing syllabus, so when she emailed the class list, she was able to discuss with her class in more detail how the exchange would proceed after receiving these letters.

On the technical writing end of the exchange, the documents students include were modified slightly. First, Guzy's students still included their resumes and wrote letters to the high school students to practice the correspondence format, but now they were able to address specific students, not only a specific audience, both in writing and responses to documents, answering students' questions and expressing encouragement about respective goals. For example, the student who received the letter previously mentioned responded as follows:

> Thank you for writing to me: it was great to hear from you. I don't follow soccer too much nowadays, but I used to watch soccer games all the time when I lived in England. My favorite English team was Liverpool. What is select soccer, anyway? I really enjoy football, and I have been a long-time fan of the Green Bay Packers. I'm glad they're doing well because they used to be stuck in a major rut.
>
> I am studying anthropology at NMSU. I'm technically a senior, but I won't be graduating for another year. Once I'm finished with my degree, I want to

get a license to teach math in the public schools. I know that math and anthropology don't seem to have much in common, but I really enjoy what I'm studying, even if I won't use it later in life. I also take the occasional math class on the side.

The anthropology department is really nice here. Many colleges require you to declare some kind of specialty or emphasis within one of the subdivisions of anthro. (The components of anthro are explained on the one-sheet.) However, NMSU allows me to dabble in whatever areas I choose. I am currently interested in archaeology and linguistics.

Are you still planning on starting at Lewis and Clark? It sounds like a great idea to me. Being close to home can make things a lot easier. My family is only a few hours north of Las Cruces, which is pretty close by Southwest standards.

There's something else that can make college easier: a really lousy summer job. I'm not talking about your everyday lousy summer job (like flipping burgers): you need to find something beyond bad. I have personally sunk as low as working the graveyard shift at a truck stop. Once you have wasted three months of your life in a job that most of your friends wouldn't touch, your college classes will seem like a breeze.

I don't have much else to say about myself. Nothing that's interesting, anyway: I spend most of my time working at NMSU's Math Learning Center and private tutoring anyone willing to pay me well. I do have a cat, though. I like dogs, but I can't truly respect any animal that can be trained to come on command. Cats are more free-spirited (i.e. spastic and cool).

I'll wrap this up now. Enjoy my resume and one-sheet. Try to use them for purposes of good, not evil (just kidding).

Second, this exchange included major changes in material; instead of revising their position papers, Guzy's students developed their skills in graphic design and visual rhetoric by creating informational one-sheets about their topics. The challenge this time was to narrow their information to the most important concepts, adjusting them to the high school audience and presenting them in an appropriate, visually interesting way. For example, the student who wrote the response letter previously mentioned produced the following one-sheet (see Fig. 1, opposite):

In our review of this second exchange, we found the process was a more successful approach for thorough audience analysis. First, the tasks we had chosen were much better suited for audience needs, and students had much better responses. Second, we allowed more time for discussion, drafting, and revising of the documents, so students could

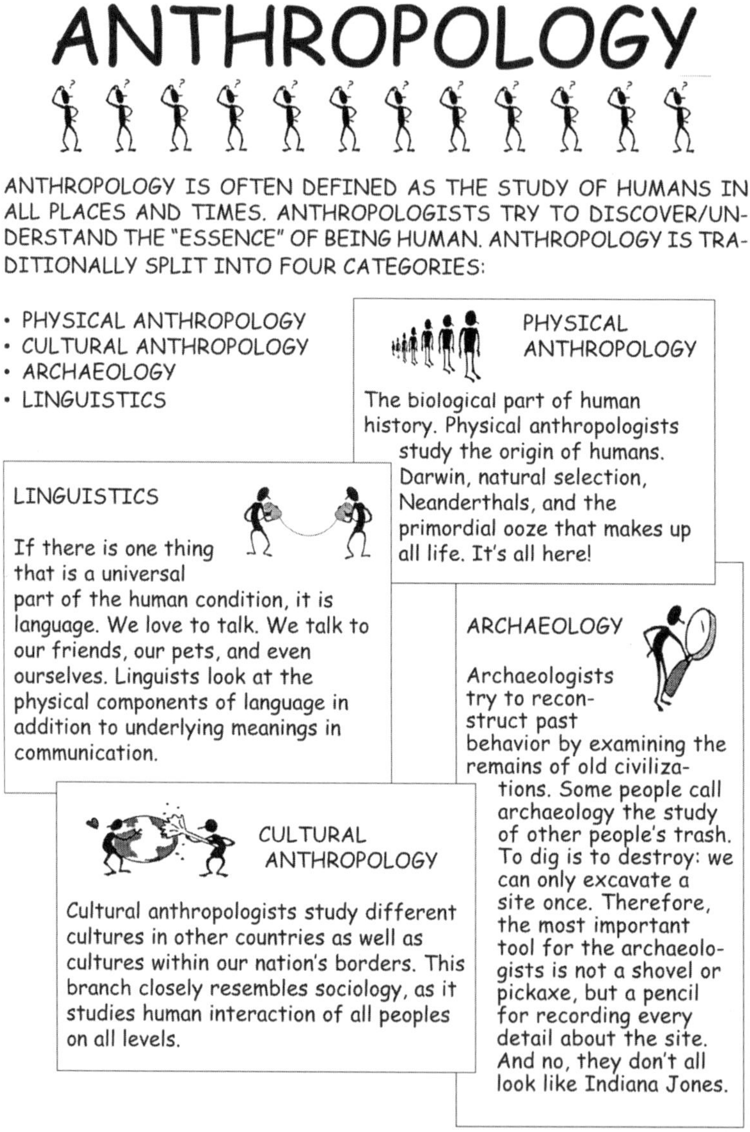

ANTHROPOLOGY

ANTHROPOLOGY IS OFTEN DEFINED AS THE STUDY OF HUMANS IN ALL PLACES AND TIMES. ANTHROPOLOGISTS TRY TO DISCOVER/UNDERSTAND THE "ESSENCE" OF BEING HUMAN. ANTHROPOLOGY IS TRADITIONALLY SPLIT INTO FOUR CATEGORIES:

- PHYSICAL ANTHROPOLOGY
- CULTURAL ANTHROPOLOGY
- ARCHAEOLOGY
- LINGUISTICS

PHYSICAL ANTHROPOLOGY

The biological part of human history. Physical anthropologists study the origin of humans. Darwin, natural selection, Neanderthals, and the primordial ooze that makes up all life. It's all here!

LINGUISTICS

If there is one thing that is a universal part of the human condition, it is language. We love to talk. We talk to our friends, our pets, and even ourselves. Linguists look at the physical components of language in addition to underlying meanings in communication.

ARCHAEOLOGY

Archaeologists try to reconstruct past behavior by examining the remains of old civilizations. Some people call archaeology the study of other people's trash. To dig is to destroy: we can only excavate a site once. Therefore, the most important tool for the archaeologists is not a shovel or pickaxe, but a pencil for recording every detail about the site. And no, they don't all look like Indiana Jones.

CULTURAL ANTHROPOLOGY

Cultural anthropologists study different cultures in other countries as well as cultures within our nation's borders. This branch closely resembles sociology, as it studies human interaction of all peoples on all levels.

Figure 1.

give more attention to their audience and their tasks. In addition, we had both seen problems with grammatical and mechanical errors during that term and decided not to make students edit these errors before sending their documents. Instead, we encouraged students to use the materials they received as editing exercises, providing an opportunity to move away from textbook exercises and brief samples to whole, "real world" samples.

IMPLICATIONS AND NEW DIRECTIONS

Throughout the exchanges, we have discussed several concerns regarding students' participation and responses to this exercise. One concern is whether we achieved true authenticity of audience analysis and response in an exercise with a secondary, although important, audience of the instructor. We have debated about whether to grade the exercise, which is an explicit acknowledgment of our participation in what ideally should be a document cycle between the students as writers and audience members. If we do grade, are we somehow influencing or "contaminating" the audience? However, without the specter of the grade looming in the background, will students take the assignment seriously and put their best efforts into writing for and responding to their audience? In actuality, we were pleasantly surprised at the growing level of enthusiasm we generally found as the exchanges progressed. Even students initially resistant to the exchange became more involved in and attentive to their writing and responses. The nature of the correspondence created almost an elementary school "pen pal" atmosphere of excitement—"Did we get anything back yet?!?"—that transcended the usual anxieties about what grades the documents would earn. In the end, we found that we did need to weigh the exchange in some fashion to ensure maximum participation, but we tended to count it more as participation than as a formal assignment grade.

As much as we want students to experience writing for these authentic audiences, recent events concerning student safety lead us to participate not necessarily as graders but as monitors of content and student conduct. In light of developments in the Columbine shooting tragedy in which a student received threatening electronic correspondence, we feel that monitoring content intended for other student readers is not censorship but a necessary measure to ensure students' academic and emotional well-being. Therefore, we discuss the appropriateness of material, especially personal information students may include in their correspondence, so as not to make their correspondents uncomfortable. If the other exchange documents, such as the informational one-sheet, contain controversial or inflammatory subject matter, such as debates over animal testing or evolution versus creationism, we draw the writers' attention to using careful rhetorical strategies in presenting the material. If resumes or letters include email addresses, we caution students about corresponding with each other off campus outside the exchange cycle for reasons of personal security and privacy. In the end, we have to remember that many high school seniors are still legally underage correspondents; if something untoward were to happen, not only would the correspondents be liable, but we, as instructors, might also have to deal with repercussions from parents and our respective administrators.

For future exchanges, we would like to take the correspondence and documentation electronic. Electronic communication would facilitate exchange of documents by eliminating printing, copying, and mailing time and costs; it would also introduce immediacy of response, perhaps encouraging extended exchanges of different types of documents. For example, we could reintroduce the position paper component on the technical communication end and give the high school students a chance for extended critique and feedback to the college students, who, in turn, could provide more thorough revision for the high school audience. We would also be able to create totally new collaborative exercises between the classes, such as how to conduct electronic searches for research paper information and how to evaluate those sources for authenticity and authority, skills students at both levels need to develop further. For these exercises to happen in a monitored environment, classes on both ends need regular access to computer classrooms, which is not currently available but should be within the next few years.

Other disciplines can experiment with this form of authenticating by arranging similar exchanges. The secondary level science and history departments can make use of cooperative learning while strengthening writing abilities. Students on both ends of the exchange could share ideas, projects, and information, aiding in a development of additional knowledge. Because writing is a key to any profession, students could develop position papers for their history classes and experiment analysis in their science classes. Drawing upon their knowledge and being forced to look beyond the classroom for ideas, these students can expand their minds and force an authentic audience to evaluate their material. At the college level, students could take on the role of master and evaluate the material, responding to material in a more authoritative manner. They may also experience a new perspective on a concept or be reminded of an old idea, but with a fresh look. Both groups benefit from an authentic audience, knowing that the material is not just a grade but a leap of faith in revealing ideas to a stranger and in the sharing of information.

Overall, we feel that this exchange is worthwhile within the scheme of each course. First, each student gains valuable experience in audience analysis by writing for a real person rather than merely completing a textbook exercise. Many also work harder on style, grammar, and mechanics to impress their correspondents. Second, through the correspondence element, students are afforded an opportunity for self-reflection at an important turning point in their academic careers. Third, by corresponding with people from geographically and culturally disparate areas, not only might they see that these people have the same types of concerns about going to school or getting a job as they do, but they also might open their minds to consider a possible future outside their hometowns. Finally, instructors can maintain some control over this type of on-campus exercise in audience analysis, not to restrict participants but to provide consistent audiences and a safe environment in which to write and learn.

APPENDIX

EXCHANGE SCHEDULE

SPRING 1997

1. New Mexico State University: Technical writing students copy resumes, revise position papers, and write cover letters for high school readers.

2. Granite City Senior High School: Senior English students review materials and write response letters to technical writing students.

SPRING 1998

1. Teachers review class rosters to assign correspondent pairs.
2. Granite City Senior High School: Senior English students write introduction letters to assigned technical writing students.
3. New Mexico State University: Technical writing students respond to high school students with a letter and an informational one-sheet that presents their position paper material.

17

ONGOING RESEARCH AND RESPONSIVE CURRICULA IN THE TWO-YEAR COLLEGE

Gary Bays

OVERVIEW

The benefits of productive collaboration between academe and industry are clear. For technical communication programs, these relationships can generate funding and new curricula. For industry, the benefits can include increased recruiting opportunities and an improved corporate reputation (Bosley 1995). Despite the inherent rewards in such partnerships, many remain sporadic and one-sided, particularly when they involve ongoing research in the workplace. This dilemma is even more pronounced in the two-year college where, historically, faculty "rarely conduct research or scholarly inquiry," yet where students are often already in the workplace and seek communication skills that will pay dividends immediately on the job (Cohen and Brawer 1989, 68).

This chapter—through a series of interviews with corporate representatives—examines a variety of issues, including how workplace research is best conducted, how industry contacts regard such research, and what corporations seek in an academic partner. At the same time, the chapter suggests that thc call for increased research coincides both with a job market that increasingly demands such research and with a two-year college student population that seeks contemporary, functional skills that translate to the workplace.

THE NEED FOR RESEARCH

In a fast-paced, global economy, technical communication programs can ill afford to conduct periodic forays into corporate culture, hoping to gather information on employee practices. Yet, according to industry representatives, select academics do just that. Researchers arrive on site to observe, interview, and survey employees, but too often the results surface only in scholarly journals far removed from the workplace—as Elizabeth Tebeaux suggests, theory that promises to "sink into a morass of verbal effluvia" with little bearing outside tenure and promotion hearings (1996, 50). Understandably, companies can grow weary of the intrusion; eventually some limit or outright deny access to their employees. As one Fortune 500 company representative who chose not to participate in this project declared, "I get a dozen requests a month to conduct research on our employees. Who am I to choose what are good projects and poor ones. It's just easier to say a blanket 'no' to you all."

Although a variety of methods for interfacing with industry have been explored at length—faculty working as consultants, as members of advisory boards, as participants in incubator centers (Powers et al. 1998; Reynolds et al. 1995; Ecker and Staples 1997)—sustained research of workplace skills merits further attention. Particularly amidst public calls for accountability in higher education, ongoing research seems a logical way to appease critics while gathering vital data to design pertinent curricula. At the same time, conducting research locally promises that curricula will reflect the needs of the companies that will hire program graduates, and documenting their abilities also plays into current assessment demands. Given the $56 billion that business and industry expend training employees annually (McCune 2000)—often on literacy issues—and polls in which companies bemoan the communication skills of new hires, ongoing research addresses a range of educational and corporate issues.

Perhaps most important, research in the workplace provides relevance to what goes on in the classroom. Research provides a measure of credibility to the classroom, particularly among students who are not technical communication majors and who too often dismiss communication courses as yet another obstacle to their chosen curriculum. Certainly, as students become increasingly critical and vocal regarding degree requirements, workplace research underscores the validity of the work done in technical communication classes. The nature and breadth of that research certainly is affected by a faculty member's available time and funding, but the act of conducting research illustrates to students that the course is immersed in the "real world"—a rallying point for the increasingly selective population in today's classrooms. Given the myriad course, program, and university selections available to students via today's technology and given the scrutiny legislators and the public increasingly aim at what happens in our classrooms, ongoing research speaks to each constituency.

For two-year college educators, though, the benefits of conducting research runs counter to their fundamental mission: teaching. Because many community college faculty members teach four and five classes and often have strict service expectations as well, research is often ignored or, as one of my colleagues remarked, relegated to "the twenty-fifth hour in the day." Still, as Cohen and Brawer found, "instructors would willingly spend more time in scholarly pursuits, as the university professors do, if they had fewer classes to meet" (1989, 142). As this chapter will suggest, with a willingness of administrators to support such research, businesses, students, and colleges can benefit and prosper from the effort.

It's important to distinguish the term "research" as it is used here. Numerous voices have expressed concerns regarding joint research efforts and their inherent issues—licensing, propriety, confidentiality, and trade secrets (Press and Washburn 2000; Lee 1998; Phillips and Metzler 1991). Rather than commercial ventures, however, this chapter focuses on research of corporate and employee communication practices and how academic research is best suited for that task. Relevant curricula and sustained relationships between business and academe rest on these. Although one vice president interviewed for this chapter suggested that "English teachers could be a little more mercenary in seeking out projects that would pay off" for them and their corporate partners, the aim of the discussion here remains on communication skills, not product development.

PARTICIPANTS AND METHODS

In establishing industry contacts for these interviews, I sought personnel who oversee research done in companies representing diverse products and cultures. The participants included here see writing generated by entry-level employees as well as by executives—a scope that permits participants unique insights into company communication skills. Eventually, I interviewed five contacts from the following four companies, representing a range of titles (director, manager, vice president) and departments (human resources, communication, research, benefits, and training).

The J. M. Smucker Company. Located in Orrville, Ohio, J. M. Smucker is known for its jams, jellies, preserves, and peanut butter. A family-run company, Smucker employs some two thousand employees worldwide and distributes products in more than sixty countries.

Russell Corporation. An international apparel company, Russell specializes in athletic wear and uniforms. The company boasts over $1 billion in annual sales and employs more than fifteen thousand worldwide.

General Motors Corporation. The perennial Fortune 500 leader General Motors employs more than 380,000 and is the world's largest automotive company, with a global presence in more than two hundred countries. The company also has more than 260 major subsidiaries, joint ventures, and affiliates around the world.

Molex Corporation. Molex is the world's second-largest manufacturer of electronic, electrical, and fiber optic interconnection products. The company employs more than sixteen thousand on six continents.[1]

The contacts responded to a series of questions exploring their impressions of academic research and business partnerships. In some instances, they did so via email; others sat down for personal interviews. In both formats, the exchanges were engaging, and the executives were gracious in devoting time and insight to the project, particularly given the time constraints they speak to in this chapter. The following questions elicited the most detailed responses.

- How do academics initiate research initiatives in your workplace?
- What are the corporate policies governing access to employees?
- Which research methods work best in corporate environments where efficiency and productivity come at a premium?
- What are the best means for sharing research findings with industry partners?
- How do you currently view academic research?
- What barriers face academics doing ongoing research in a corporate environment?
- What writing/communication issues merit research in your workplace?

Making Contact

According to those interviewed, efforts to initiate academic research in the workplace rival Baskin-Robbins for variety. These include the following strategies:

- Cold calls
- Contacts made through training and consulting
- Contacts made through advisory board memberships
- Formal proposals for on-site research
- Face-to-face meetings at professional conferences
- Requests handed down from CEOs and CFOs
- Informational interviews
- Longstanding university-business partnerships.

In the case of the last item, for example, General Motors maintains a strong working relationship with both the University of Michigan and Michigan Technological University, while Auburn University allies itself with Russell Corporation. The benefits of these relationships are invaluable because partners are already working with known quantities.

Unlike the scenario depicted earlier in this essay, none of the participants here denied my request for an interview; however, most of them emphasized the need for personal contacts as a means of initiating partnerships. As one interviewee noted, "You need to develop relationships in advance of asking for favors." Accordingly, respondents felt cold calls and unsolicited proposals offered the least promise for initiating partnerships.

The principal issues regarding access to employees, understandably, are confidentiality and privacy. Contacts admit being wary of releasing any personnel information—job titles, addresses—without a clear explanation of how that information will be used. In some instances, companies enforce stringent privacy policies that direct such activity; and human resources personnel, legal counsel, or review boards can be involved in approving research in the workplace. Clear research proposals or abstracts of the proposed research are vital to company officials who must secure approvals. Academics are well advised to work up the description of the project before seeking permission to do it. On completion of the work, companies can also make explicit demands that employee information be destroyed. Heeding these instructions certainly goes a long way toward developing a working relationship.

The corporate culture also influences how companies react to research requests. At the J. M. Smucker Company, officials admit being "very cautious to protect" employees. The company underscores its family orientation in all of its corporate literature, and fourth generation Smucker family members currently direct the company. When interviewed as part of *Fortune's* list of the "100 Best Companies to Work For," Chairman Timothy Smucker claimed the company continues to build on the "basic beliefs and ideals instilled in the company" a century before, among these, "the value of people" (Plauche 2000, par. 3). Without that background, a zealous researcher might unknowingly tread on values the company holds dear. The lesson seems clear: before entering the workplace to conduct research, academics must study and understand the environment in which they will be working.

In marked contrast, gaining access to more than 380,000 employees, like those at General Motors, can be an exhaustive and time-consuming process not only for researchers but also for company officials. When conducting a recent survey of employees at General Motors, I submitted the parameters of the study (college graduates, corporate environment, first or second year on the job) to corporate officials who then designed a program to identify recipients. Because the company conducts hundreds of research efforts internally each year, my request was prioritized and put in the queue for consideration. The wait was exacerbated by contract negotiations during this same period that demanded the attention of employees involved in this project.

Clearly, without the full cooperation of patient industry contacts, such information is simply unattainable, and researchers need to court their partners with this in mind. Making demands in such partnerships based on academic calendars holds little sway in the workplace. This issue promises to be less critical once academics establish a level of trust in a partnership, but in an initial project researchers must respect and abide by the constraints governing their research in the workplace.

Gathering Information

Once permitted access to employees, academics must design research tools likely to elicit both valid and timely information. On this point, interviewees had the most divergent suggestions:

Written surveys or questionnaires (at desk, at home, proctored)
Telephone interviews
Focus groups
Email questionnaires
Workplace observation

In each case, researchers must recognize the hackneyed reality of the workplace: time is money. According to one corporate representative,

"Time is the key issue; the bigger the demands of the research, the less likely employees will take the time to complete it." Another interviewee noted that any research method requiring "more than fifteen to twenty minutes" will limit a wide response from busy employees. Given the range of methods suggested by corporate contacts here, it's worth examining the relative merits of each.

Certainly, surveys and questionnaires seek answers to specific questions—often with a specific range of responses available to respondents—and these tools are quite efficient in gathering material. If they are well designed (see Anderson 1985), surveys can easily be completed within the fifteen- to twenty-minute time frame suggested previously. Survey responses may vary depending on whether the questions are completed at work, where employees may feel obligated to complete them, but may not feel free to express critical opinions, or are done at home, where employees may express themselves more freely, particularly if their responses are guaranteed anonymity. However, without the subtle pressure exerted by the workplace, employees may completely disregard questionnaires sent to their homes.

Telephone interviews invite more elaborate responses from employees, permitting a far broader discussion than surveys offer. Like surveys, however, they come with the same drawbacks depending on whether they are conducted at home or work. Too, researchers must devote much more time to actually speaking with employees and transcribing the responses. According to one respondent, focus groups offer extensive feedback because employees can compare their comments with those of peers. The research method comes as a double-edged sword: focus groups take numerous employees away from their work simultaneously, and in corporate environments where productivity comes at a premium, companies may well avoid projects using this method of research.

Perhaps the most efficient form of feedback today comes by way of electronic mail. Given its speed and relatively painless demands for responding, email allows researchers near-immediate data; too, there are fewer issues of confidentiality involved compared to releasing home addresses.

Workplace observation is a mixed bag for both researchers and business partners. For academics, it permits a way to gather information without demanding employee time. In "shadowing" an employee, researchers can collect accurate information on time devoted to communication acts, tools and methods of communication, and strategies for creating and revising documents. One respondent endorsed this method enthusiastically. "It's a win-win for both partners. You get information on how employees work, but you don't stop them from *doing* the work." For academics, the drawbacks include limited numbers on which to base research findings and significant time demands required in shadowing employees. The power of such observation is borne out by previous research (for example, Selzer 1983), but obviously two-year college faculty already facing time constraints may find this type of research out of the question without substantial administrative support.

Sharing Findings

As noted earlier, workplace research often serves as valuable evidence for promotion and tenure hearings, but companies may get little benefit from the academic findings. If published, the work is relegated to scholarly journals, far removed from the eyes of participants, in formats and language equally distant. Clearly, this dynamic is hard to characterize as a "relationship." At the same time, given the frenzied pace within their respective companies, contacts appear to have little time for repeated conversations with academics.

Thus, there was consensus among those interviewed about how best to share our research with them: executive summaries. Whether it was a summary attached to a "brief report" or a summary of "no more than a page and a half or two pages tops," contacts emphasized the need for brevity in reporting findings.

The working relationship, of course, is also defined by the corporate culture in which the work is conducted. For example, General Motors has its own corporate university, a trend that has seen marked growth during the past decade (Meister 2000; Michaels 2000). Research that fits the pedagogical aims of such in-house efforts may help academics make inroads with corporate partners. Indeed, these inroads may lead to "exciting and lucrative relationships for colleges" (Meister 2001, B10). In contrast, officials at smaller companies like the J. M. Smucker Company admit they react to training and research needs as they become aware of them. As a respondent from that company noted, "We are need-driven; when we hear of an issue that requires attention, we dedicate time and resources to it."

Of course, this need should motivate colleges and universities to maintain an ongoing relationship with local business and industry; understanding and tailoring research to the "needs" of their partners is one way to maintain that relationship in addition to more traditional means (conducting training workshops, consulting, and so forth). When academics fail to sustain such relationships, the risk is twofold: (1) researchers will miss vital opportunities to work collaboratively through company needs, and (2) our graduates will miss out on valuable opportunities to develop skills that address those needs. Although advisory boards and committees drawn from local business and industry can lend anecdotal evidence to college programs, first-hand research provides far better evidence on which to build those programs. Given that many two-year college students are already in the workforce or likely to enter the job market within the year, providing a relevant *and* current curriculum is an issue of immediacy. It is fair neither to the companies that hire our graduates nor to the graduates themselves when academics remain unmindful of workplace skills and deficiencies.

Pitfalls and Perceptions

Academics also need to enter research partnerships with a keen understanding of the barriers they are likely to face in the workplace, and respondents to this survey described those barriers in detail. The first is simply getting in the door. As illustrated earlier, companies are inundated with requests for research—and often they simply deny all requests. Respondents expressed concerns about legal and organizational practices publicized by academic research. For example, one industry representative noted her company would be far more inclined to okay research about communication practices companywide than research of management or executive skills, particularly if the latter put the company in a bad light. Respondents noted that depending on the findings and the way in which they are publicized (journal article, conference presentations), as a matter of professional courtesy, the company should get a copy of the work. Most agreed potentially critical research might come under much closer scrutiny by those responsible for approving the work. Finding the niche between genuine academic inquiry and corporate agendas might be the most difficult task in initiating workplace research, and it is an issue that merits further discussion among technical communication professionals and business representatives alike.

It's no surprise that interviewees agreed on one barrier to employee research: time. According to one respondent, academics must design "research that will not be a drain on people's limited time or the limited human resources" available. The relevance of the research also plays a vital role in gaining access to employees. As one respondent noted, "the more germane the research, the more likely the cooperation from the company and the employees themselves. Those involved need to know 'What's in it for me?'" This becomes increasingly important as companies, in the name of efficiency, downsize and assign more responsibilities to those remaining on the job.

Although the "disconnections between industry and higher education" have diminished through increased interchange (Ecker and Staples 1997, 380), the disconnections nonetheless remain. Though politic in their comments, it is obvious that select members of business and industry still see a schism in the relationship. One respondent didn't "feel academics look beyond their rarified air—they're not forced to stay abreast of the world outside the Ivory Towers." Another admitted there is "probably a communication gap between industry and academia in making that [research] translation," while another, using the language of the issue, asserted, "There is a disconnect between what businesses need to solve as problems and nebulous research studies that can't be easily translated to solving business issues." As Rehling (1998) argues, the notion of the "one-way exchange" in which business shares privileged information with academics remains, and academic researchers and corporate partners must consider how they can develop truly collaborative exchanges.

Continuing Research

Respondents suggested and agreed on a number of communication issues deserving attention from workplace researchers in the future.

- Grammar issues
- Brevity and succinctness
- Presentation skills
- Analysis of written materials
- Electronic communication
- Corporate politics and ethics

Although admittedly a condensed research agenda, nonetheless it provides academics with ideas for continued study of workplace practices.

By sharing ongoing research needs and subsequent findings, college and industry partners can work collaboratively and build what can be characterized as "relationships."

The Road to Tenure

Although the merits of research are relevant to the classroom and to the community, faculty must increasingly look to research for their professional survival. As a recent article in the *Chronicle of Higher Education* (Wilson 2001) explained, "The bar for tenure is rising at major research universities and teaching institutions alike. Most departments demand more published research—either articles or books, or both." In higher education today, faculty recognize the role research plays in securing a position and in eventually seeking tenure and promotion. According to Wilson, the trend is not solely for those teaching at research institutions: "research is now a key factor at many institutions that previously focused almost entirely on teaching and service (A12). Two-year colleges are benefiting from the new wave of scholarly colleagues entering its ranks, but without ample administrative support (released time, research funds) few of those colleagues will likely sustain their research in light of heavy teaching and service expectations.

CONCLUSIONS

Research that leads to relevant curricula has been an ongoing theme in technical communication literature (Tebeaux 1996; Ecker and Staples 1997; Rehling 1998). In this chapter, I suggest how that research might be initiated and best conducted, based on feedback from corporate representatives. The relevance of workplace research speaks for itself: it appeases critics who call for "real world" curricula; it promises to assess the outcomes of what educators do in the classroom; and it brings credibility to classroom instruction in perhaps the most competitive era of student recruitment.

Academics who undertake a research initiative also face a fundamental issue that merits further professional dialogue: How can industry and colleges support academics facing daunting teaching loads and funding issues as they undertake workplace research? Even on a small scale, research is labor intensive, particularly when it includes comprehensive research of a corporation and its culture, as suggested here. Some institutions are seeking a humane balance in their quest for heavier research productivity. The University of Richmond, while making greater research demands on its new hires, also offers a progressive approach to helping them produce research. Teaching loads have been reduced in recent years, and the School of Arts and Sciences just instituted semester-long leaves to support research efforts among junior faculty (Wilson, 2001). This idea and others merit attention; it is incumbent upon administrators and industry to support faculty, who now face higher research demands.

The means to do workplace research may well decide the success of the research itself. As Hayhoe (1998) suggests,

> Technical communicators in the academy and industry need to explore a new model of education for the next millenium, one that fosters, promotes, and actively pursues learning—and learning to learn. Only by discovering our own limitations and collaborating effectively with those whose strengths complement our own will we truly enrich ourselves, our students, and our audiences. (20)

The call for ongoing research may also make us reevaluate what constitutes meaningful research. Sullivan and Porter (1997) note the schism between traditional or empirical researchers and those working from newer perspectives. The work described in this chapter draws on a relatively small number of participants, yet the findings offer means of closing the gap between academe and business. Combined with other studies of comparable aims, the work adds to a growing body of workplace research that can enrich lives as Hayhoe suggests. As I note in this chapter, establishing a true relationship with a business partner demands that academics work within the constraints of those partners; that in turn promises to change the nature of how research is conducted.

My experience collecting work for this project speaks to the enrichment Hayhoe describes. A number of students from my college find employment with the J. M. Smucker Company upon graduation each year; so, I sought out my institution's director of workforce development for a contact within the company. As a result of one hour-long interview and a subsequent survey of their first-year employees, I eventually revamped elements of my technical communication course. The survey suggested that employees felt particularly vulnerable writing summaries, and yet summaries ranked high among the types of writing they were required to do each day. Thus, I introduced additional summary writing into my course, as well as more public-speaking opportunities, another skill employees noted among their daily demands and deficiencies. I also found that most of the first-year employees learned about the company and its writing practices from their immediate supervisors. Thus, my next project with the J. M. Smucker Company will involve a survey of supervisors and their impressions of the writing they encounter daily in the workplace.

The time involved in this work—including survey design, mailing, analysis, completion of an executive summary, and the initial interview—reached thirty to forty hours spread out over a series of months, and the findings became part of a panel presentation at the Conference on College Composition and Communication. The research led to a subsequent interview with an influential senior executive, and we plan to meet on an annual basis to discuss workplace communication issues and, potentially, courses and programs that will serve the needs of students and the company's employees.

Still, I was able to maintain close contact with only two of the five contacts I made at the outset of this work—a result that speaks to the limitations imposed by a full teaching load and the service expectations of my college. Certainly, working with a local company proved far easier than those at a distance and, ultimately, far more relevant to students. The concept of "taming the hydra," that is, balancing teaching and scholarship, has been addressed before (Andrews Knodt 1988). However, it is time that two-year colleges both recognize the importance of research to their futures and, in turn, support faculty undertaking such work. It is work that must not be relegated to the "twenty-fifth hour" of our days.

Ultimately, the parallels between academic and business partners are striking. Both hope to be productive; both put a premium on time; and both must spend their respective resources judiciously. Given the speed at which communication practices change in business and industry, technical communication programs can ill afford to ignore the vital role of ongoing workplace research. Those efforts, while goaded by increasingly rigorous tenure guidelines nationwide, must be tempered by realistic expectations of faculty productivity. Ongoing research remains vital to the future of technical communication programs. As we attempt to design relevant courses for students in our classes, we must do so with equally relevant research that builds on truly collaborative efforts between colleges and industry.

18

EXTREME PEDAGOGIES

Teaching in Partnership, Teaching at a Distance

Billie Wahlstrom

OVERVIEW

There's something to be said about the adrenalin rush found in extreme sports—bungee jumping off cliffs, parasailing from city buildings, solo canoeing down the Noatek River, twenty-six miles above the Arctic Circle. The desire to try something risky and survive, to push oneself and redefine one's identity, to go boldly forward seems to describe one part of human nature. Without the explorers and adventurers our maps would be blank, our sense of human potential would be limited, and we would have no epic sagas to tell around our campfires or tales to tell our grandchildren.

Too bad that the sense of adventure doesn't permeate the academic world. Here our maps are pretty much blank, our sense of human potential is limited, and being bold generally works against getting tenure. Nevertheless, this is the time for extreme pedagogies, and this is a call for them. Technical communication programs, classes, and teachers need to change, and the change being called for isn't modest. Taking risks where there is only a thin margin of safety isn't for every program, administrator, or faculty member, but it's essential for those programs and people who have glimpsed the future, see the opportunities, and can stand the adrenaline rush.

Most of us have a hard enough time keeping our classes up-to-date, our programs running smoothly, and our lives in order. The idea that we might need to make fundamental changes in our courses, transform our programs, and rethink ourselves isn't very appealing. We say things are going well enough, but the trouble is that things aren't going well enough for technical communication. We've been in a series of steady and unresolved debates over the last twenty years about what we are, what we value, how practitioners and academics should work together, and what the real focus of our research should be. Meanwhile, the typology of the world—with its learners and its technology—shifts beneath our feet.

The analogy between extreme sports and extreme pedagogies isn't just something fanciful. As is true for our culture as a whole, business as usual in technical communication pedagogy and programs won't work much longer. The climbs and drops of the Dow and the Nasdaq independently and relative to each other show us the rough ride we're in for while the business and investment communities try to understand what is happening as commerce and new technologies intersect. Commerce and technology aren't the only two areas of society colliding. Technology and education—even our fundamental idea of what a university is—have collided as we move from the analog model to the digital one. Put simply, we need bold approaches because "American education is structured around a technology [books] 500 years old. We have reached the limits of that technology" (Ellsworth, Hedley, and Barbatta 1994, 59).

In the 1990s, we heard a call for transformation of higher education. From renowned neuroscientist Michael Gazzaniga (1998) we heard that "it is time to reorganize the whole university," and, from our technical communication colleague Ann Hill Duin (2001), we are urged to participate in the technologically driven transformation of the learning marketplace. The voices calling for change in the academic world have been so loud there is now a counterchorus singing the refrain of "leave it all alone." Yet, it's too late for that, and what was a call for change a few years ago has become a discussion of what those changes actually look like (Hanna 2000).

Certainly, much is working well in our field. We have more programs, textbooks, book series, journals, and students than in the past. About four hundred thousand students take technical writing classes each year, and technical communication has been added to the graduation standards for high school students in several states, including my state of Minnesota. The numbers of practicing technical communicators has increased steadily, and *U.S. News and World Report* (Avery et al.1998) singled technical communication out as one of the twenty hottest jobs of the decade. Yet, the nature of education is undergoing well-documented transformations, and the world of knowledge making, information storage, and digital technology is changing how we do our business. Much as we might like to leave things as they are, technical communication teachers, administrators, and researchers need to cope with and shape the changes underway. Few models exist for what's possible, and I'll propose one example of what might be done and point to a few additional possibilities. But the map is largely unfilled in, and what matters more than following anyone else's lead is addressing the issues boldly and without delay.

What I describe here is a model for technical communication programs of the future that began developing at the University of Minnesota in the 1990s, in partnership with three other institutions. And, although I use this program as one example, what I propose is not based on a single program at any institution. Rather, this is a call for extreme pedagogies that transform the local technical communication programs into multisite, multimodality, multiaudience, and multipurpose initiatives. And it's a discussion about what such extreme pedagogies engender—sites of resistance and sites of acceptance and support—as well as an analysis of where current models fall short.

DIFFERENT LEARNERS AND DIFFERENT NEEDS

Our first job as educators is to think about who students are. As technical communicators, we need to apply audience analysis to this question. Adult learners, seeking work-related education, make up the vast majority of people seeking knowledge and information these days. In a presentation on distance education and the internet that futurist Michael Dolan made at the University of Minnesota in 1999, he pointed out how significantly the learner had changed from what most of us in the academy were trained to consider. For one thing, degree and certificate programs are no longer the primary site of education for adults. In 1998, although more than 45 million people were engaged in work-related learning, only approximately 11 million learners were pursuing traditional educational goals: certificates, degrees, and credentials.

Most technical communication programs do not directly serve this population; the number of programs actively engaged in credit and noncredit continuing education is very small. Most of our programs are academic, and as a consequence, we have relinquished the education of the majority of technical communicators to others—the companies that employ communicators and educate in-house or independent contractors who offer job training in the field. At my institution we offer a B.S., an M.S., an M.A., and a Ph.D. in scientific and technical communication. Aside from scattered events for alumni and occasional programs, no systematic or strategic plan exists for keeping our graduates current with continuing professional education.

This is a serious mistake for our programs and a loss for society as a whole. Certainly, we have prepared our graduates to keep up-to-date by informing them of resources such as professional organizations and publications. Moreover, many of our graduates are in research and development roles in their organizations and know more than we do as their teachers about issues of practice and innovation.

Nevertheless, we have a critical role to perform, and we have given the task to others. Not all of our graduates have ended up in the same industries, and one function we should be performing is providing systematic education on new trends and their social and design implications. Let me provide a concrete example. How do we expect our graduates who went through our print-focused programs over the years to think about and work on design issues for talking cars and the small screens? In 2001, Gartner Research predicted a 260 percent increase in unit sales of PDAs, from 8.39 units in 2000 to 33.7 million in 2004. Everyone from the Navy to students and doctors in medical schools use these (*Bloomberg News* 2001). The issue isn't simply one of how do we design for low resolution screens—160 x 160 pixel resolution—instead of the resolution of a standard computer monitor at 640 x 480, a reduction of resolution of 1,200 percent. Articles in *Technical Communication* (Kim and Albers 2002) and other journals can provide an overview of these issues and provide design guidelines.

What an article cannot do, however, is bring our practicing professionals up to speed on what these changes mean and how to think about them. How do human constraints on memory and attention, for instance, combine with technological limitations of communication media to present serious problems in how data are presented and interpreted in these new situations? At the University of Minnesota, for example, we are considering making radiological films available for PDAs and laptops wirelessly. Do the problems in resolution for handheld devices mean that interpreting these films online might result in users missing important data? If we screen a mammogram on a PDA, might we miss a tumor? Or as the number of "multitasking" cars increases—from 5.4 million in 2002 to 28 million in 2005—how will we think about attention (and what detracts from it) as we design new message systems? Research from the Progressive Auto Insurance Company and the University of Iowa shows that using cell phones and listening to voice-activated email negatively affect braking time and our ability to stay in our lanes (McCafferty 2002). Physicians have a basic standard of at least doing "no harm," but without systematic continuing education for our graduates—backed up by ongoing research in our academic departments—I don't think we can make this claim.

The nature of the learner isn't all that has changed; how the learner is getting access to education is changing as well. As Dolan pointed out, there were one million online learners in 1998. Each year there are more learners because more educational opportunities are available online. For example, at the University of Minnesota, the very popular Master Gardener Program has sprouted the Internet Master Gardener Program for an even wider audience.[1] But we don't see a similar pattern in our technical communication programs. Because the average college education is out of date within five years, the great majority of practicing technical communicators will need continuing education, but we are thinking still about degrees and certificates that require their presence on our campuses.

I'm not suggesting that technical communication programs abandon the students who come to us to take classes, but that we meet the needs of both practitioners and students who do not intend to or cannot drop everything and come to campus; we must develop other solutions. Michael Markel (1999) put it succinctly: "Technical communication students are looking for the convenience and flexibility offered by distance instruction" (209).

Why are established programs so slow to use the Internet for distributing course content widely? Certainly, we're not afraid of digital technology or putting materials on the Internet. A search of the Internet using Google in March 2000 turned up results of about 20,100 entries for "Technical Communication" in .75 seconds, and by June 2002, such a search turned up 103,000 hits in .16 seconds. Information proliferates and access increases, but what is revealing is that most of these sites are commercial, advertising books and workshops in a variety of delivery modes. Of the academic sites, most advertised traditional on-campus programs and conventional courses offered in traditional classrooms. They have some version of the statement provided by the Professional

Writing and Technical Communication Program at the University of Massachusetts Amherst: "Although the program features an extensive Web site with a great many online resources, we are classroom-based and do not offer courses online for distance learning."

Only a few of these sites were for programs fully online or with substantial online components and a well-developed Internet presence. Most notable among those are Texas Tech University (http://english.ttu.edu/tc/), which offers a master's degree in technical communication via the Internet, and the New Jersey Institute of Technology (http://www.njit.edu/MSPTC/), which offers an M.S. in professional and technical communication online. Many academic programs offer only a smattering of courses online or on ITV.

Technical communication programs use the Internet extensively for a range of purposes, but we are well behind the curve in terms offering our educational expertise in what Ann Hill Duin (2001) calls the emerging "learning marketplace." Our progress in meeting the needs of learners by offering our expertise via the Internet is slow for a number of reasons—technological, administrative, and personal. At the University of Minnesota, for example, we have a certificate in Professional Communication, but for a student not on the Twin Cities campus to figure out how to complete this program requires persistence beyond what is required of on-campus students. The program description reveals the difficulties students enrolling have to face:

> Courses are offered through a variety of methods, including on-campus classes scheduled days and evenings and selected courses available through interactive television, online, or correspondence. . . . Many of the on-campus classes are offered late afternoons or evenings so that they are convenient for adult learners (registration in daytime classes is also available). Courses carried on interactive television will be available simultaneously to students at the University of Minnesota-St. Paul campus and in classrooms at the University of Minnesota-Crookston, the University Center in Rochester, and at Southwest State in Marshall, MN. Online and correspondence courses have no classroom meetings and can be taken from any location. Online courses require Internet access and skills. (See http://www.cce.umn.edu/certificates/cert_org_prof_comm.shtml.)

Obviously, this program isn't a learner-centered or even a user-friendly model. Whether the failure to address student needs grows from organizational inertia or resistance, research suggests that before an organization will change "there must be enormous external pressures" and there needs to be "a plan, model, or vision" (Hanna 2000, 29).

From the inside trying to change a university is "like trying to move a battleship at rest with one's bare hands" (Weinstein 1993). Funding, as well as political, individual, and institutional issues, would be manageable if we had a collective commitment to a bold vision of meeting the needs of all learners in our field and employing the technologies necessary to serve them well. We remain too focused on traditional, place-bound courses and degrees to see how the new economy and academic partnerships are restructuring education.

In his essay, "How to Change the University," Michael Gazzaniga (1998) speaks of how narrow our vision of education has become and the result of this provincial outlook:

> Someone once said that Americans think the weather starts in California and ends in Maine. We all have such a limited view of our existence. Intellectual pursuits change, as does everything else. In today's world, trying to maintain a personal garden is an endeavor that will die on the vine. (237)

Gazzaniga proposes transforming the University by removing the restrictive partitioning of ideas caused by disciplinary lines and reorganizing the institution in ways that reflect modern intellectual life. Extreme pedagogies help us shed the restrictive partitioning of our programs in individual institutions and our restrictive vision of who students are and where they should be learning. Individual programs, narrowly construed curricula, and the physical walls of the classroom are the personal gardens that we cannot sustain.

PROPOSING THE PARTNERSHIPS

Technical communication programs can no longer flourish in isolation. Extreme pedagogies call for technical communication programs to do three things simultaneously:

- Work in partnership with other institutions to craft technical communication programs from their shared resources
- Offer their curriculum in innovative combinations of degrees, certificates, and noncredit forms to meet learners' needs
- Deliver their content in a variety of modalities

It's not cost effective, wise, or ethical to duplicate technical communication programs in a region. Technological innovations that removed the barriers of time and space for students have also removed them for institutions, although many of our colleges and universities haven't quite figured out what to do with that freedom yet. Within a region, it makes better financial sense to establish partnerships that allow community colleges, four-year institutions, and universities to cooperate on meeting learners' needs.

A Partnership Model

The idea of a technical communication program forming partnerships is nothing new (Karis 1997). Many programs have industry partners and other partners within various departments in the university. What I'm proposing are partnerships among technical communication programs. This multiprogram model is one that we've tried at the University of Minnesota. The rhetoric department at Minnesota offers a certificate in Professional Communication; a B.S. and an M.S. in Scientific and Technical Communication; and an M.A. and a Ph.D. in Rhetoric and Scientific and Technical Communication. We have about sixty-five undergraduates in our program and a like number in our M.S., M.A, and Ph.D. programs together. For about ten years we have had 100 percent placement of students within six months of graduation. There are about ten jobs and internships for each student, although not always in exactly the area a student would like to pursue. We are supported by our college administrators and have adequate, although not extraordinary, resources.

Launching an extreme pedagogy like multiprogram partnerships isn't something one does without prodding. Ours came from the college office, which asked all its departments to undertake partnerships as a reaffirmation of the university's land-grant mission. College administrators identified state institutions for partnering and provided seed money to facilitate visits to campuses and planning. Consequently, our B.S. in technical communication is now offered in St. Paul and in partnership with three other institutions, with which we found a natural fit:

1. A joint degree with University of Minnesota at Crookston—a coordinate campus three hundred miles to the northwest
2. A degree program with Rochester Community and Technical College (RCTC)—seventy miles to the southeast—that has students taking two years' work at RCTC and the other two at Minnesota, via distance offerings
3. A joint degree program with Southwest State University, a university in a different state system, three hundred miles to the southwest

Forging partnerships at the University of Minnesota took years of good-faith negotiations, visits, and conversations. Creating a different memorandum of agreement for each site took substantial department, college, and university-wide efforts. And even though some of the problems are still being worked out—who has access and what kind to the libraries of the University of Minnesota system and who keeps what student records, for example—the results are far better than we could have imagined. Students are enrolling, internships are developing, and recruiting is underway throughout the state.

Minnesota's partnerships mean that students can remain at their home institutions and receive a University of Minnesota degree without ever setting foot on campus. They have access to local resources and the resources of their partners. All students take specialized core courses offered from the University of Minnesota and complete the degree with courses partners agree are equivalent across sites. These partnerships make the partners' resources available throughout the region, allow each partner to craft the degree for the needs of its students, raise enrollments everywhere, and increase social capital by uniting many people behind a single project.

What does this model suggest beyond Minnesota's limited experience? First of all, it suggests that traditional universities can use various or multiple strategies to extend their programs. Programs can duplicate themselves at other locations using "resident campus faculty who travel physically . . . or who teach electronically to one or more locations" (Hanna 2002). This approach allows for successful courses to be offered at multiple sites, combined with courses developed specifically at each site. Or if they wish, programs can develop a successful offering with a stable curriculum and set of requirements at one site and then recreate it in multiple locations, using different faculty to teach at each location. This approach allows for a uniform knowledge base, with identical courses offered across all locations. These approaches can work when a region has a single program with the resources to reach out.

A Common Pool of Shared Resources

The next question is what do we do when there are several programs in a region, but no one of them can or wishes to step forward to extend its traditional program? Not all institutions have the same resources available to them, and the "advent of ubiquitous networking technology will lead to the centralization of key functions in the education system," as we are seeing in business (Schank 2000, 44; see also Duin, Baer, and Starke-Meyerring 2001). Pooling support structures and supplies is one place we can see a significant advantage for technical communication programs. Technical communication programs are expensive to staff because of software, hardware, and equipment needs. Pooling resources can allow us to purchase software, for example, less expensively under a site license that covers all. Unlike the other models discussed, which extend individual programs, this model is aggregative in nature.

In addition to supplies, partners can pool infrastructure expenses. For example, if one institution has an online writing center that provides asynchronous tutoring by trained tutors to all visitors, it is possible to join with partnership institutions to share resources. Technology allows us to create a seamless and customized interface and what Schank (2000) called a "centralized pool of tutors" who could serve all users and not "dilute the brand" of any institution, as seems to be a fear of potentially competing institutions (45).

Nurturing the Partnerships

Partnerships need nurturing because our inclination is to look at each other competitively. The central site in the partnership can take the initiative for bringing partners together regularly by fax, ITV, and email, as well as for face-to-face visits. At St. Paul, we have a program staff person with the primary responsibility for coordinating the partnerships at our end. Faculty, staff, students, and program administration all need to be nurtured. Faculty, for example, at the University of Minnesota partnerships meet together twice a year to go over curriculum, troubleshoot, and discuss technology needs. Students need to meet with teachers from the different sites and with administrators so they have the opportunity to say what works and what doesn't. At one such meeting with students, for example, faculty received generous praise for their availability. Students liked how the teachers "got right back to them" using fax, phone, and email. They felt part of the class. Investment in regular meetings allows staff to coordinate offerings, calendars, book orders and to build the personal relationships that carry these joint endeavors over the rough spots.

Tangible and less tangible rewards can be had from partnerships. For one thing, each partnership offers something different. For example, at the University of Minnesota, support from one of our partners resulted in shared funding to support a faculty member with a joint appointment at our institution. At another time, support from our program enabled our partner to lobby for an additional faculty member to help with the partnerships and increased student interests. In partnership settings, students gain by finding job opportunities and internships in places where they might not otherwise find them. For example, students at a partner institution identified internship possibilities at a large frozen-food company and at corn-processing plants that students at another campus would never have discovered. There are increased opportunities for grants and coauthored papers. And lastly—but not to be minimized—is the opportunity to make new friends and find new colleagues. When a faculty member at one of our partnership institutions developed cancer, the concern at the other sites was genuine, as was the joy we shared when another colleague had a baby and emailed all the partners her good news. Professionally as well, the opportunity to write letters of support for tenure and promotion as well as to find partners for conferences are important returns on investment for faculty.

INNOVATING IN CURRICULUM DESIGN

Even with innovative partnerships, the focus of education at Minnesota and elsewhere in the country is on the traditional courses and degrees. This is a serious limitation of this model we've designed and one we hope others will address. Our extreme pedagogy isn't extreme enough to meet the needs of all the learners we must serve. To serve learners, technical communication programs and teachers must rethink curriculum from the ground floor and must integrate revising the curriculum in the context of the technologies we can use to enhance and deliver it. John Chambers of Cisco Systems gives us a sense of the change we face in our programs: "The next big killer application of the Internet is going to be education. Education over the Internet is going to be so it is going to make email usage look like a rounding error in terms of the Internet capacity it will consume" (Duin, Baer, and Starke-Meyerring 2001, 1).

To make the curricular changes needed and to thrive in the technological environment in which we find ourselves, we must value flexibility far more than we do. We have to discard the notion that education must be packaged into fifty-minute classes offered three times a week. We must rethink the units of curriculum, how they're delivered, and to whom.

Different "Classes"/Different "Programs"

For the last ten years, many voices in technical communication (Selber 1997; Hawisher and Selfe 1999) have been telling us that "technological and curricular change" are not independent (Werner and Kaufer 1997, 313) and that we must rework our curricula. Unfortunately, most of our response to these calls has been the production of scholarly essays discussing change rather than engaging in actual programmatic change. Tiffin and Rjasingham suggest that we ask, "What kind of system is needed to prepare people for life in an information society?" (1995,1) and not assume that what we're currently doing will work tomorrow. To get us to think more broadly about the sites where education will take place in the future, they no longer use the word "classroom" and urge us to focus on what they call "virtual learning spaces" (10). And, importantly, they emphasize the home as a site where new technologies enable learning to take place.

A major component of extreme pedagogies is creating those virtual learning spaces in our technical communication programs. We cannot stop with putting the traditional curricula online. Yet, if the majority of people are adult learners needing education for the workplace, then we should examine the knowledge assets held by each of our programs and think about ways to combine and distribute those assets to reach more people. For example, when we approve a new "course" in our curriculum, we should expect it to serve in multiple roles:

1. As a traditional course to meet the needs of residential students
2. As an online, streaming video or ITV course to meet the needs of nonresidential and partnership students seeking degrees and certificates
3. As a series of knowledge pieces that can be repurposed into credit and noncredit modules and made available to a wide range of learners in a variety of media—from face-to-face to CD-ROM to the Internet.

Fortunately, the development of Learning Content Management Systems (LCMS)—not Learning Management Systems, which manage learners and track their progress rather than manage learning content—is making this last vision a reality. LCMSs based on a learning object model allow us to combine and reconfigure learning materials to meet individual learners' needs. An LCMS, as Brandon Hall describes it (<www.brandon-hall.com>), allows educators to create, store, reuse, manage, and deliver learning content from a single object repository.

LCMSs are developing quickly, allowing for authoring, content assembly, object repositories (that can be coordinated with libraries), delivery engines, and multiple output formats.

In this emerging environment of reusable learning content, faculty's intellectual property must be protected while we carry out our goal of reaching the broadest possible set of learners. This protection can be done in two ways. First, some new LCMSs include intellectual property tracking software that notes each time a "learning object" representing an individual faculty member's proprietary material is used, gives credit, and, if contractually determined, calculates the amount due to the information's creator. Secondly, we can explicitly include faculty rewards in our discussion of revising curriculum. Many faculty consult to supplement their income and to keep themselves aware of trends in the workplace. By broadening our notion of "curriculum" to include noncredit, continuing education materials, we offer the chance for faculty to "consult" within our programs and to profit from their intellectual efforts.

This process of modularizing a class and repurposing the knowledge pieces for various audiences and technologies is something my colleague Paul Brady and I have done with some of our courses; and it has resulted in a tremendous alteration in the way we teach and think about "course content." One technical communication course with which we have done this is Message Design II: Theory and Practice. This course was created originally for delivery face-to-face for fifteen weeks at the rate of two class meetings each week. Before putting it online, I performed an information audit of its content.

When I rethought my class into its knowledge components, here is some of what I found:

1. Forty-five units of specific information—such as interactivity, linear and nonlinear information structures, designing complex messages, participatory design
2. One hundred twenty readings along with reading guides—such as book chapters, material on the Internet, essays, clippings, newspaper articles
3. Fourteen assignments—such as educational materials, a public service announcement, a report on media effects, a provocative question
4. Forty-five class activities—such as class critiques, brainstorming sessions, games, simulations
5. Thirty discussion questions for the online chat

6. Two hundred ten commentaries on class assignments, one comment per assignment per student
7. Ninety related Web sites to visit—such as the Yale Style Guide for multimedia and Bobby, a site for improving accessibility of Web pages to people with disabilities

Once I imagined my class in components other than the fifty-minute session, I had a very different idea of how those components might be recombined to meet the needs of different students. Paul Brady, our college coordinator of Instructional Computing, designed and built for me an e-library with a Web interface—complete with stacks and private holdings—in which to store the knowledge pieces that made up my class. From this collection of basic information, I am able to pull specific learning materials for different users: an online class with practicing professionals, the residential class with degree seekers, and noncredit workshops for practitioners in the workplace.

I did this rethinking of my course as a personal response to essays I've been reading on the need to repurpose intellectual materials. Realistically, however, most faculty need more tangible rewards to go to this trouble. Faculty need a reward for reworking their materials for an online or ITV environment, and if they are willing to share some of their materials with other colleagues at their home institution as well as with partners, they need a return on their investment. When a faculty member's efforts result in tuition revenue from students truly new to the system, then providing a piece of the action, as well as rewarding innovation in the department, is appropriate. Without this plan, faculty will take their expertise and ideas to the for-profit institutions in their digital neighborhood and reap the rewards where they can.

Different Programs

One place where the Minnesota model has failed is that we have no plans to develop courses together from pooled intellectual assets, despite our partnership in offering a degree together. Programs that look at their overall course content as flexible information assets can decide to separate and recombine those assets to reach more learners. If departments think not about whole courses, but about the information components of all the courses in the curriculum, we are free to combine information from a number of courses to put together—with relative ease—whole new components.

For example, many courses in the technical communication programs at the University of Minnesota have a small unit on usability testing. Combining all those components and adding the reading guides, resources, and assignments associated with those components could produce useful results. First, students could draw on a variety of materials to understand usability testing. Students unfamiliar with the concept could draw on beginning materials, and students who had experience could tap the advanced materials, regardless of which class they found themselves in. Secondly, the complete holdings could be combined to create (with the help of a faculty facilitator, whose efforts would be recognized financially or otherwise) noncredit units for people in the workplace who want to become familiar with this testing. If several programs pooled their information assets, whole new "courses" could emerge.

Different Classrooms

Genuinely rethinking curriculum forces us to reconsider the sites at which learning takes place. I'll leave the ergonomic issues to someone else, as well as the discussion about connectivity, media source switching, and the available equipment (electronic whiteboards, laser discs, wireless standards, document scanners, VCRs, monitors, and computers) for the new "connected classroom" (see, for example, Coppola and Thomas 2000). What's more useful is to think beyond even the wired or wireless classroom to the virtual learning sites mentioned earlier.

A large enough body of theory and practice exists for us to work successfully in these e-spaces—from Lynnette Porter's (1997) and McCormack and Jones's (1998) practical guidebooks on creating e-classrooms and Web education to studies helping us understand how to meet distance learners' needs of social and academic integration (Eastman 1998). Today's body of scholarly literature can help us build successful online learning spaces. And there is enough now on how to keep our focus on pedagogy in general and on technical communication pedagogy in particular rather than spending all our energies on thinking solely about the technology (Knox 1997; Jorn, Duin, and Wahlstrom 1996). Indeed, for years our technical communication colleagues have been writing about our need to blow out the walls of the technical communication classroom (Allen and Wickliff 1997) by establishing a variety of collaborative projects that bring students close by and at a distance together in a common learning space. We've just not carried their suggestions to their necessary conclusions.

PROPOSING A MIXED MODALITY

Just as partnership make good sense economically and educationally, so does offering courses in mixed modalities. Not everyone learns best from having an instructor lecture. Learners bring with them a variety of learning preferences—from aural to tactile. If learning styles make a difference in the ability of learners to master material, then we can better meet learner needs by offering our materials variously formatted.

Internet-based education is ideal for this. Increasingly, at the University of Minnesota we are building audio, video, interactivity, and opportunities for interacting with others into all our courses. We started out doing this for the benefit of our online students but quickly discovered that the materials designed for them worked just as well for traditional, in-class students. These days we are using enhanced online syllabi for our residential classes so that we can link students on campus to the body of resources originally created just for the distance students. In fewer than five years we went from no classes and no students accessing Web CT for course-related materials online to more than forty-one thousand students systemwide doing so in one semester of 2002.

The question is how to convince teachers who have used only the chalk and talk method to think in new terms. Many faculty seem to share Sven Birkerts's (1994, 199–201) perspective, lamenting the passing of the primacy of the book:

> Everything has changed in the past quarter century, with the changes hitting their real momentum in the past decade or so. . . . But even these shifts [in the publishing business] . . . are as nothing to the real transformation, which is that of the cultural context. There is no denying that a terrible prestige-drop has afflicted books themselves. They have moved from center to margin; the terms of their mattering are nothing like they used to be. You do not have to be a writer, a publisher, or a critic to see this. Anyone who pays attention knows that writing and reading are not what it's about these days.

Many faculty—who have successful careers largely because texts alone work best for them as a learning option—share Birkerts's distrust of moving beyond the book to teach:

> The gathered concentration of [reading] is no longer our central cultural paradigm. . . . Reading is taught, of course, and books are assigned in school, but any teacher you ask will tell you that it is getting harder and harder to sell the solitary one-on-one to students. The practice itself is changing. Already it

> is clear that the new reading will be technology-enhanced. CD-ROM packages are on the way . . . to gloss and illustrate, but also to break the perceived tedium of concentration by offering Interactivity options, and the seductions of collage-creation. Don't just crack your brain on Hamlet, but pull up pictures of famous actors and directors and read some sidebar interviews, even view clips of scenes in performance.

To dismiss the opportunity to teach literacy in media beyond print seems a terrible arrogance on our part, and to privilege one technology when using others can provide more people with access to knowledge seems reactionary. Technical communication pedagogy isn't the academic equivalent of the History Channel. Fundamentally, extreme pedagogies are about meeting the needs of the learner rather than the teacher.

Finding the Right Mix

Extreme pedagogies don't mean that everything within a technical communication program must change. Programs must revise their mix of offerings and technologies for delivering them, however. The mix in each program will depend on its partners, its mission, and its faculty. At Minnesota, for example, finding the right mix resulted in our agreement that we would not offer a Ph.D. online. Faculty decided that too many resources and social interactions were missing online, and they felt that the seminar experience is essential for the students we wished to turn out. Other programs will need to make their assessments.

The Cyborg Teacher

A critical component of extreme pedagogies is the teacher. The Internet creates a "twenty-four/seven" learning environment in terms of the availability of content to the student. Those numbers don't describe the availability of the teacher to the student, obviously, but it is foolhardy to think that new ways of organizing and delivering content don't seriously change the role of the teacher. Technical communication instructors old enough to remember teaching before email can already see how learning spaces are changed by technology. Technologies that "allow users to share everything that can be sent over the Web" and that are "highly collaborative" are certain to affect teachers more extensively (Wahlstrom 1997, 142). Adapting to this environment will call for technical communication teachers to develop their Cyborg selves to be comfortable and successful in this new environment. It's critical to remember that online education is education foremost, and issues of teaching skill remain central. As Arbaugh (2001) points out, many challenges faced online are the same as those faced in the classroom. Extreme pedagogies cannot succeed unless we provide teachers with support in learning how to "exploit the communicative and adaptive capabilities of the new technologies" in the building of communities of learners (Laurillard 2002).

Training the New Technical Communication Teacher

In 1999, Linda Clemens, then an M.A. student at the University of Minnesota, was concerned about the teacher training future Ph.D.'s were receiving in technical communication programs. She wondered how well these new professors would be prepared to function in distance delivery classes and how skillful they would be in making use of new technologies in all of their work. She argued that "technical communication Ph.D. students must come to understand the attributes of each type of technology to support their effort to select technology appropriate to the learners and to the purpose of the course, and to support their efforts to merge instructional design, teaching strategies, and technology" (119).

To see how well our future teachers were being prepared, Clemens examined what technical communication scholars were writing about these new pedagogies and what teacher training was being offered in the Ph.D. programs in our field. Her results weren't heartening. In the first place, she found that "we have no distance education theory specific to technical communication" (123). Moreover, aside from a single course at one of the Ph.D.-granting programs, there is no specific training for future Ph.D.'s in technical communication that focuses on distance delivery and the uses of new technologies in the classroom. Training the "next generation of the professorate" to teach well in the digital classroom is a subject of discussion at many universities, but it has not received the attention it needs in our discipline.

SITES OF RESISTANCE/SITES OF SUPPORT

Resistance

Despite the most compelling arguments from our field and from the academy in general, there is still much resistance in technical communication and elsewhere to the pedagogical changes we must make. Not every program is ready to bungee jump—no matter what the potential for personal growth. A call for extreme pedagogies is being met with resistance—not from everyone, but from many places. Resistance is natural and even helpful, but it cannot be allowed to prevail.

Resistance of Faculty

Finding technical communication faculty a site of resistance is understandable. Technical communication has often been marginalized in the academy. Our colleagues in English sometimes minimize those of us who study workplace writing instead of the well-crafted novel or current literary theory. Technical communication teachers have generally turned to critical studies and postmodern theory to build their credibility in the academy and have written frequently and angrily about the pressures put on them from industry to teach functional literacy and skills (Wahlstrom 1997). In many ways new pedagogies, because they involve working with new technologies, seem at first to be a capitulation to tools rather than a focus on ideas.

Secondly, working with different technologies than those with which you were trained and upon which you based your academic career is bound to cause conflicts and resentments. The problems of teaching with ITV and the Internet are well known and there is much discussion of faculty feeling technology as a constraint rather than as a tool to enrich what goes on in the classroom (Johnson-Eilola 1997). We cannot minimize these issues, but we can take a proactive approach, such as that described by Racine and Dilworth in this volume. Teaching with technology and using the Internet for education are here to stay, although change will occur both rapidly and extensively.

The challenge in working with faculty resistance is to involve faculty in the discussion of the technology and to work to prepare them to use it well, to be successful. Equally important is the need for discussion about the new pedagogies we must develop and how we will "preserve our pedagogical convictions," as Racine and Dilworth (this volume) put it. Yet, fundamentally, we must acknowledge that we cannot continue with business as usual. Stuart Moulthrop (1999) put the issue succinctly: "Turning toward the past does not excuse one from the present" (419).

Rewards

I've described some of the rewards we have set up for faculty willing to practice extreme pedagogy at the University of Minnesota, but we need to think beyond established faculty to new faculty as well. The answer isn't to protect new faculty from participating in our efforts to reshape our discipline. Rather, it is to reassure new faculty that this investment will pay off with merit increases, teaching awards, tenure, and publications, and to see that it does. Classroom research has been an important area of scholarly work in our field. These new pedagogies expand the technical communication classroom, and much needs to be written about how we tame this wild space. As Linda Clemens (1999) pointed out, we need to develop the body of theory dealing specifically with technical communication and distance delivery, and this is a genuine opportunity for new faculty to publish work badly needed.

Resistance of Students

Students also resist new technologies and new pedagogies that include partnerships. There is student resistance both locally and at our partnership sites, for example. At home there is just the hint of jealousy and a competition for attention. Local students say that if we didn't have ITV courses, for example, they would have us all to themselves and wouldn't have to deal with students they don't know. Additionally, students may have concerns about access. If their schools don't have universal email, for example, they may wonder how they can reach the teacher when they need help.

Student resistance must be addressed and overcome if all learners are going to feel valued and content. Much of this resistance disappears when students are given a chance to discuss the reasons partnerships were established and the limitations and advantages of various technologies. Helping technical communication students grow past their resistance is relatively easy. For one thing, we expect them to be able to communicate effectively using the materials at hand (to paraphrase Aristotle), and today's materials include interactive television. We can tell students honestly that the opportunity to explore ITV, for example, will give them an advantage in the job market. They can learn about how to make presentations on ITV that will provide them invaluable experience for the corporate world where more and more interactions among people at distance sites are mediated.

Secondly, technical communication students need to know how to work in groups, and these days collaboration is often mediated. Making the classroom a learning site where students learn the content and also learn to communicate effectively using the medium at hand is a useful approach. If we don't address their frustrations, then students at a distance secretly turn off their mikes so they can talk to each other and not to the teacher. But students can be engaged, as Dilworth and Racine discuss in this volume, by being part of a conscious process in the classroom of claiming the medium.

Resistance of the Academy—The David Noble Factor

Change would be a lot easier if we were able to point to a collective will for change among technical communication faculty and administrators. Without the presence of positive models, people resistant will turn to dissenting voices.

In the face of technological change and despite the need to reach all learners, the David Nobles of the world see it as their task to defend the "sacred space" of classroom and the old ways of doing things even when there is no attempt to abolish all traditional means of education. As Young (2000) pointed out in the *Chronicle of Higher Education,* "David F. Noble says distance education is fool's gold, and he's eager to point out who the fools are" (A47).

Noble is a strong critic of technology, who "refuses email, . . . often writes drafts of his books by hand" and "certainly doesn't have World Wide Web pages for his courses." In 1997, he wrote *Digital Diploma Mills,* which began his attacks on distance education. His critiques deal with how higher education is selling itself by commodifying information and establishing what he calls unholy alliances with commercial interests. The result he says leads to a disenfranchisement of faculty. He argues that the commitment to technology and its infrastructure is "a technological tapeworm in the guts of higher education" (Young 2000, A48). His criticisms—who owns what and who is served by decisions about technology—need to be taken seriously, but they shouldn't stop us in our tracks.

Locating and distributing alternative perspectives is helpful in keeping the marketplace of ideas from becoming a monopoly. Ben Shneiderman, director of the human-computer interaction laboratory at the University of Maryland at College Park, helps put Noble—and the perspectives he represents—into context: "His fear-filled rhetoric and whipping of the boogie-monster of entrepreneurial corruption of education is misleading, shallow and even counterproductive" (Young 2000,

A48). Because resistant colleagues are given to forwarding Noble quotes via email to faculty involved in distance and new technology initiatives, it's useful to know Noble's arguments. The answer to this resistance is, in part, dialogue between the resistor and the adopters. And in part it's keeping up-to-date on the research on quality and effectiveness of distance education offerings and programs. For every David Noble, there is an Ann Hill Duin (Duin, Baer, and Starke-Meyerring 2001) and a Donald Hanna (2000), who articulate reasons for optimism and change.

Acceptance and Support

Not surprisingly, students and faculty are sites of both resistance and support. For students at a distance, the chance to interact with peers at other universities is fun, especially if they are not put in a position where their first encounters are intimidating or designed to compare student populations.

Feedback from students in a variety of e-classrooms at the University of Minnesota indicates that their experiences are not unlike those of teachers and students elsewhere. For example, the students in an e-classroom at Pace University found the ITV environment a "no-doze classroom" when run by committed teachers. As the Pace students put it, they found the room "'professional' and they felt as they must come to class prepared and ready to work" (Coppola and Thomas 2000, 34). In my role as vice provost for Information Technology at the University of Minnesota, I get at least one email a week from students who take the time to let me know how an online class enabled them to continue their programs while taking a semester abroad or when they were at home in bed with a tricky pregnancy or when knee surgery made coming to campus impossible.

Feedback from partners is also encouraging. Many faculty at other institutions feel, as we do, that pooling our resources means we can all do a better job serving the needs of students. Their support is apparent in their calls and email and in the fact that the personal as well as professional are covered in our discussions. Partnership faculty are committed to the project model and the community building we tell students these technologies enable. Faculty who have participated in the distance programs have, generally, been enthusiastic. When faculty have not found the process rewarding, it has often been because of technological difficulties or differences in the cultures of the partnership places.

One problem is getting faculty who are unhappy to let someone know about the problems; too many suffer in silence to avoid appearing inadequate before their peers. In general, we've found that when we find out about problems we are able to fix them. There is a general sense that people involved in these programs want them to succeed. Despite that, untenured faculty may feel that they are jeopardizing their future if they raise concerns or have difficulties. Developing ways for faculty to raise concerns in a protected environment is a significant component of a successful program.

Another problem that we hesitate to discuss involves the faculty who bring their bad habits to the electronic environment and find them magnified and then blame the technology and undermine department efforts. What do we do with the faculty member who doesn't update material, doesn't answer email, doesn't allow for any interaction in a course? The answer, in part, is that we uphold the standards of good teaching and require that everyone be evaluated on a common set of standards—availability, promptness of response, ability to inspire, amount of useful information provided—not materially affected by technology at all.

WHAT OF THE PRESENT AND WHAT OF THE FUTURE?

The Commodification of Information

The question is what future do we want and are we determined enough to go get it? We have to consider David Noble's reservations about the commodification of information and the risks of establishing partnerships that place technology before pedagogy. The issue of the commodification of information he points to is a problem for technical communication programs because this commodification limits access. Basically, we're in the literacy business, and our job is to make sure that the university—in whatever format it appears—is a place where unbiased and robust discourse can take place.

Meeting the Needs of More Learners

If we don't provide the education needed in technical communication to those asking for it, then for-profit companies will step right in. For-profit companies are taking over a lot of education, particularly in our field. Online Learning is a good example of the for-profit organizations providing courses and programs to students. It offers a program in technical communication through its business and management offerings:

> OnlineLearning.net, a Sylvan Learning company, has accepted more than 20,000 enrollments in 1,700 online courses since 1996. Accredited, graduate-level extension courses are offered in teacher education and business and management. In recognition of its superior service and quality content, the National Education Association selected OnlineLearning.net as its partner in online education for teachers.

Given the changing nature of technical communication practices, it seems foolish if we give over the continuing education of practitioners or the education of degree seekers to others when we, in fact, are the sources of new knowledge in the field, and we are public institutions, invested in the public good.

CONCLUSION

We simply cannot continue to do business as usual. Soon there will be no business worth doing if we keep to the established paths. Where are the innovators and risk takers in our profession and in our discipline? There is no map to show us where to go; we will need to explore, to venture down rivers that undoubtedly contain rapids. On this adventure, I, for one, need to recall frequently what the dean of my college told me when we were discussing curriculum and the future: "Whatever you do, make sure that it's bold."

ENDNOTES

NOTES TO CHAPTER ONE (DUBINSKY)

1. Barton and Barton (1993) describe how European mapmakers controlled the way most individuals saw the world. For example, they describe how, by making Europe the center of the world, the mapmakers made the continent of Africa, which in reality is considerably larger than Eurasia, appear smaller.
2. I could argue, and would accept the argument, that all service is valuable. Certainly, from a religious perspective, any act freely given (washing feet, spinning cloth) that contributes to the well-being of others is valuable. However, the distinction I'm making has to do with those acts that do not come voluntarily but as the result of the condition of employment.
3. See David Russell's (1991) discussion about how most disciplines see writing as the elementary skill of "talking with the pen instead of the tongue" (quoted on page 6). Although necessary, it isn't usually deemed worthy of sacrificing time in their classes to discuss.
4. Ernest Boyer (1981) describes this connection in detail in *Higher Learning in the Nation's Service,* a manifesto of sorts, arguing for a return to an environment in which the ideal of service takes precedence over seeking knowledge for knowledge's sake alone.
5. The focus on scholarship occurred almost simultaneously with the increase of enrollment due to the Morrill Act.
6. Campus Compact had grown from three institutions in 1985 to over 477 institutions by 1994 (Cha and Rothman 1994).
7. In his *Politics,* Aristotle asked, "Should the useful in life, or should virtue, or should the higher knowledge be the aim of our training?" (Book 8, sec. 2, 1905)
8. In my article, "The Ideal Orator Revisited: Service-Learning as a Path to Virtue," I outline how I shifted from a pedagogy that didn't balance service and learning to one that did. By using student evaluations and reflection journals, I illustrate how, regardless of the strategy, well over 90 percent found the service project the most valuable. However, I also illustrate how, by shifting my emphases and corresponding course

materials to balance service and learning, students began to see the importance of their work in terms of building reciprocal partnerships with their organizations and working with them to effect change.

NOTES TO CHAPTER FIVE (GRABILL)

1. Certainly technical and professional writing has exploded because of its work in "non-academic settings." But to this point, those settings have typically been white-collar workplaces and a rather narrow range of professions. This focus is understandable and has been productive, but it need not be the only focus. My interests here are in work in community-based settings, a phrase that describes a diversity of contexts, writers, and audiences, all of which involve technical or professional writing of some kind.
2. Of related interest should be the work of service-learning practitioners in composition, who often configure their service in terms of writing for nonprofits, which is clearly professional, if not also technical writing. In this respect, service learning in technical and professional writing is more widespread than I am presenting it here. At the same time, one could challenge this practice within composition (and with first-year writing students) as one who calls into question the identity of the first-year writing course and the ability of first-year writing students to be of service to nonprofit organizations.
3. By sophisticated and writing related I mean that the projects have a clearly defined problem—it is clear that a problem and therefore a project exist—but the details of the problem need further articulation, the problem itself requires research to be successfully addressed, and the solution to the problem involves writing of some kind. For technical writing classes, the problems I look for are typically "technical" in some way (for more on this topic, see Huckin 1997, 52). But when I do projects in classes such as business writing, I am open to a wider range of possibilities.
4. My students certainly have some choice. They have options as to which classes they take and need not take a service-learning class. They have choices as to which projects they work on, and I usually accommodate their preferences. In some instances, I have been able to present multiple project possibilities before I have made any commitments and therefore created a space for students to help me choose projects that were most meaningful to them in terms of both social/community issues and the research and writing involved.

5. My evidence here is almost completely anecdotal, and so I am more than willing to admit I am wrong. In fact, I hope I am. But as I read and participate on listservs, read articles, and listen to papers and talks given at conferences, I am fairly convinced that my characterization is accurate.
6. By "involvement," I mean discussions between service providers and clients about issues of policy—not, for instance, about issues related to individual care. So, for example, broader policy discussions would involve issues of what adolescents affected by the disease needed and wanted from service providers or the council itself, not the issues individual adolescents were dealing with at the time (such as particular physical, social, or psychological problems or needs). The line between the two, sometimes, disappears.
7. The Atlanta Project was an initiative spearheaded by The Carter Center, who saw in Atlanta problems they often observed in third world countries (for example, poor health care) and sought to address these issues by using techniques often successful in third world contexts—supporting community-based and grassroots efforts (as opposed to delivering a program from the "outside"). The Office of Data and Policy Analysis (DAPA) began in 1992 as a partnership between the Georgia Tech City Planning Program and The Atlanta Project. DAPA serves as a planning advocate, data intermediary, and information warehouse for agencies involved in neighborhood planning and community development in Atlanta.
8. One reason that the configuration of the tool begins with these features is that the current clients of DAPA, the same people who will still be primary users of this Web site, are comfortable with data maps and the analyses associated with them.

NOTES TO CHAPTER SEVEN (BRIDGEFORD)

1. Many technical communication teachers, for example, use the Challenger disaster as a context for discussions about ethics. The problems described in this case study have long been identified as a failure of communication between the manufacturers of the O-rings and NASA engineers. The context this case study offers has been so well defined, it has become overdetermined in such a way that students don't need to figure anything out. The narrative voice in the case study identifies the problem and the solution for them. Because the limited perspective of this case study is told in such as way as to

identify for students the nature of the communicative problem, students don't need to think, to figure out what went wrong, or to determine what answer they should conclude. The answer is built into the problem—it is controlled. Because communicative decisions occur concurrently, that is, holistically and comprehensively, they require interpretative acts. In this way literature provides a complex, comprehensive "situation" that must be interpreted before an answer can be offered. A narrative way of knowing, I argue, provides a basis for interpretation, encouraging students to draw on their background knowledge, to use their minds in ways already familiar to them, and to act accordingly: in essence, to think for themselves.

2. The best answer I received to this procedure occurred during a semester in which I used both *Terrarium* and a case study about Torch Lake in the Upper Peninsula, which is identified as one of the Environmental Protection Agency's Areas of Concern. A student wrote, "*Terrarium* is the exigency of the EPA."
3. For the purposes of this study, the stories were limited to print narratives (either short stories or novels), although in later courses, I opened up these choices to include other narrative genres such as poetry, movies, song lyrics, and cartoons.
4. Student writing analyzed in this chapter is printed verbatim as students submitted it. I have made no changes to punctuation, grammar, or spelling.
5. In this class, students completed projects individually.

NOTES TO CHAPTER EIGHT (KALMBACH)

1. See also Walsh (1977), Hogan (1983), Moran and Moran (1985), Trace (1985), McDowell (1987), Charney and Rayman (1989), Tovey (1991).
2. Sherer (1984), for example, tells of an employer who often takes home between 150 and 200 resumes a night. One shudders to picture this individual reading resumes during dinner while calming screaming children or between scenes of *ER*. If a resume doesn't catch him or her quickly, it doesn't have a chance.
3. See Nemnich and Jandt (1999) and Krause (1997).
4. For discussions of HTML resumes, see Nemnich and Jandt (1999) and Quibble (1995).
5. I do not require a separate scannable resume because at this point in time too few students benefit from redoing their resume solely to put

the resume in scannable form. In central Illinois, only one major employer uses Resumex; and for students who do not have a complex background, a carefully designed print resume can also serve as a scannable resume. In the near future, more students may benefit from separate print and scannable resumes, but for now I am more comfortable with the scannable resume as optional component of the assignment.

6. State Farm has a Web page with advice on creating resumes that can be easily scanned at http://www.statefarm.com/careers/resprep.htm.
7. One alternative to missing fonts and incompatible file formats is to create an Acrobat PDF version of a resume with embedded fonts. Unfortunately, although Acrobat may be the most reliable way to electronically transmit documents designed to be printed, PDFs cannot be easily moved into resume databases. They are a complement, not a substitute, for other forms of the resume: another form of hypermediation.
8. Saving to the Web does, however, have its own issues. Students too often publish material on the Internet without reflecting on the consequences of that act. They may include their phone number and address in an HTML resume without thinking about the implications of making this information available to anyone. Teaching the HTML resume also means discussing these ethical and practical issues.
9. Even Bill Gates adds his two cents worth in the form of the Microsoft Word templates many students use as a starting point for their print resumes.

NOTES TO CHAPTER TEN (KITALONG)

1. See, for example, http://wired.st-and.ac.uk/~donald/humour/misc/compidiot.html and http://www.Webgasm.com/DanoSays/IdiotProof.html).
2. See, for example, statistics available at http://www.nua.org/surveys/how_many_online/index.html.
3. John Lannon's popular technical communication textbook employs a rare tripartite system in which users are classified as technical, semitechnical, or nontechnical, but the effect is the same—purification.
4. This remains true even though Microsoft's tactics are often regarded as somewhat controversial.
5. In a more innocuous reading, the sheep represent network packets, and the bridge represents an inadequate access line.

NOTES TO CHAPTER ELEVEN (ZERBE)

1. Herndl's (1993b) critique is a notable exception.
2. This approach can be adapted to other areas of scientific and technical communication. For example, instructors could ask students to compare traditional scientific treatments of a particular subject with those from sources disparagingly referred to as "pseudoscience" or "junk science" outlets.
3. By no means do I want to imply that scientists and physicians and other health professionals *intend* to ensure that humankind does not attain perfect health. I wish only to point out that, for example, despite the fact that a vaccine for polio was developed in 1955 (polio itself was identified in 1916), the AIDS virus HIV was already at work; some estimates show HIV in action as early as the 1930s. AIDS itself was not formally identified until 1982.
4. By "signifying practices," I mean "material effects of language in the conduct of human affairs" (Berlin 1996). Berlin does not use this phrase to define "signifying practices" explicitly, but I believe that it captures the idea that language produced by cultural institutions has a real impact on individuals—who, it must be pointed out, can also use language that may become a signifying practice. Later in the paragraph, in the Myers quote, I interpret "signifier" to refer to the institution responsible for producing a particular sign (that is, a textual or visual portrayal of, in the sense that Myers discusses, human beings or human institutions) and "signified" to refer to the people whom the sign is designed to represent.
5. The drawbacks most often cited (Karras 1999) are the following:

   - Mammograms from younger women are much more difficult to read and interpret than those from older women because younger women have denser breast tissue.
   - More false positives and false negatives occur in mammograms from younger women.
   - Women in their forties who obtain annual mammograms may develop a false sense of security and stop performing breast self-exams.

6. Because numerous and disparate therapies and other activities comprise alternative and complementary medicine, the idea that alternative and complementary medicine is a cultural institution may be debatable. However, Lutz and Fuller (1998) point out that alternative and complementary medicine, as a whole, is deeply involved in a legitimation

process that may lead to institutionalization. Indeed, Mowbray (2000) has demonstrated how descriptions of acupuncture in Western medical discourse have become much more positive in tone over the past several decades; this discursive acceptance is associated with recognition of acupuncture in Western medical practice—as evidenced by the establishment of an Office of Alternative and Complementary Medicine (in 1992) at the National Institutes of Health.

7. The URL for this Web site is http://cancernet.nci.nih.gov/cancer_types/lung_cancer.shtml#toc2. Click on "patients" under "Treatment of Non-Small Cell Lung Cancer."
8. The URL for this Web site is http://www.tianxian.com/English. I asked students to look primarily for lung cancer information under "Testimonials" and "Tian Xian Products."
9. The students with whom I piloted the questions were enrolled in my undergraduate 200-level Introduction to Technical and Scientific Communication or in my graduate 500-level Scientific Rhetoric courses.

NOTES TO CHAPTER TWELVE (SELFE)

1. In the rest of this chapter, I'll use technical communication (TC) to refer to courses and curricula from programs that include this name and professional communication. I do this not because I think the distinctions between such programs are inconsequential but because the assignment/method I suggest in this chapter can easily be adapted to many English studies disciplines: composition, literature, English as a second language, and so forth.

NOTES TO CHAPTER THIRTEEN (MEHLENBACHER AND DICKS)

1. See http://www2.acs.ncsu.edu/UPA/planning/spdr_956.htm.
2. See http://www.ncsu.edu.
3. See http://www.ncsu.edu/server_statistics/1998Sep20.html#Archive.
4. See http://www.nsf.gov/sbe/srs/seind98/access/c7/c7s4.htm#c7s412.
5. See http://www.acm.org/technews/articles/2000-2/0324f.html#item2.
6. See http://www4.ncsu.edu/~brad_m/gradsymp.html.

NOTES TO CHAPTER FOURTEEN (HANSEN)

1. I am writing from some experience in these areas. For almost ten years, our technical communication B.A. and M.S. programs have incorporated service- and client-based projects at all levels in the curriculum, working with small businesses and, through Metropolitan

State's Center for Community Based Learning, a wide variety of nonprofit organizations.

2. An ideal internship seminar would meet at the beginning of the term (to introduce students to the responsibilities and realities of an internship), one or more times during the course of the term (to report progress and discuss challenges/successes), and, perhaps most importantly, at the end of the term—a final meeting where students summarize and reflect on their internship experiences. I provide some structure to this final meeting by asking students to respond to a series of prompts. The students (and I) greatly enjoy these meetings: they become very engaged in each other's experiences (especially when it involves problem solving). This type of seminar might be easily accommodated to the Internet.

NOTES TO CHAPTER SEVENTEEN (BAYS)

1. Backgrounds on these corporations are at smucker.com; gm.com; russellcorp.com; and molex.com.

NOTES TO CHAPTER EIGHTEEN (WAHLSTROM)

1. See http://Webct3.umn.edu/public/HORT/1003demo/.

REFERENCES

Adams, E. J. 1993. A Project-Intensive Software Design Course. *SIGCSE Bulletin* 25:112–16.

Allen, J. 1989. Breaking with Tradition: New Directions in Audience Analysis. In *Technical Writing: Theory and Practice*, ed. B. Fearing and W. K. Sparrow. New York: MLA.

———. 1992. Bridge over Troubled Waters? Connecting Research and Pedagogy in Composition and Business/Technical Communication. *Technical Communication Quarterly* 1:5–26.

Allen, N., and G. A. Wickliff. 1997. Learning Up Close and at a Distance. In *Computers and Technical Communication: Pedagogical and Programmatic Perspectives*, ed. S. Selber. Greenwich CT: Ablex.

Almstrum, V. L., N. Dale, A. Berglund, M. Granger, J. C. Little, D. M. Miller, M. Petre, P. Schragger, and F. Springsteel. 1996. Evaluation: Turning Technology from Toy to Tool. In *Integrating Technology into Computer Software Environments Conference Proceedings*. Barcelona: ACM P.

Alred, G. J., W. T. Oliu, and C. T. Brusaw. 1992. *The Professional Writer: A Guide for Advanced Technical Writing*. New York: St. Martin's Press.

Althusser, L. 1971. *Lenin and Philosophy and Other Essays*. New York: Monthly Review Press.

Anderson, P. 1985. What Survey Research Tells Us About Writing at Work. In *Writing in Nonacademic Settings*, ed. L. Odell and D. Goswami. New York: Guilford.

———. 1995. Evaluating Academic Technical Communication Programs: New Stakeholders, Diverse Goals. *Technical Communication* 42:628–33.

Andrews Knodt, E. 1988. Taming Hydra: The Problem of Balancing Teaching and Scholarship at a Two-Year College. *Teaching English in the Two Year College* 15:170–74.

Anson, C. M. 1999. Distant Voices: Teaching and Writing in a Culture of Technology. *College English* 61:261–80.

Arbaugh, J. B. 2001. How Instructor Immediacy Behaviors Affect Student Satisfaction and Learning in Web-Based Courses. *Business Communication Quarterly* 64:42–54.

Aristotle. Politics. 1941. *The Basic Works of Aristotle*. Trans. Ed. Richard McKeon. Benjamin Jowett. New York: Random House

Artemeva, N., S. Logie, and J. St-Martin. 1999. From Page to State: How Theories of Genre and Situated Learning Help Introduce Engineering Students to Discipline-Specific Communication. *Technical Communication Quarterly* 8:301–18.

Asimow, M. 2000. Bad Lawyers in the Movies. *Nova Law Review* 24.2 (winter). Available online at http://tarlton.law.utexas.edu/lpop/etext-/nova/asimow24.htm#92 [cited September 28, 2002].

Avery. *U.S. News and World Report*. 1998. Best Jobs for the Future. 84–85.

Avery, D., M. Charski, D. Floyd, M. Loftus, M. B. Marcus, A. Mulrine, S. Schultz, and K. Terrell. 1998, October 26. "20 Hot Job Tracks" *US News and World Report*.

Bacon, N. 1997. Community Service Writing: Problems, Challenges, Questions. In *Writing the Community: Concepts and Models for Service Learning in Composition*, ed. L. Adler-Kassner and R. Crooks. Urbana IL: NCTE.

Baker, M., and C. David. 1994. The Rhetoric of Power: Political Issues in Management Writing. *Technical Communication Quarterly* 3:165–79.

Barton, B. F., and M. S. Barton. 1993. Ideology and the Map: Toward a Postmodern Visual Design Practice. *Professional Communication: The Social Perspective*, ed. N. R. Blyler and C. Thralls. Newbury Park CA: Sage.

Bellah, R., R. Madsen, W.M. Sullivan, A. Swidler, and S. M. Tipton. 1985. *Habits of the Heart.* Berkeley: University of California Press.

Bellotti, V., S. Buckingham Shum, A. MacLean, and N. Hammond. 1995. Multidisciplinary Modeling in HCI Design. In *Proceedings of CHI'95 ACM Conference on Human Factors in Computing. Boulder*. ACM P.

Berkenkotter, C., and T. N. Huckin. 1995. *Genre Knowledge in Disciplinary Communication: Cognition/Culture/Power.* Hillsdale NJ: Lawrence Erlbaum.

Berlin, J. A. 1993. Poststructuralism, Semiotics, and Social-Epistemic Rhetoric: Converging Agendas. In *Defining the New Rhetoric*, ed. T. Enos and S. C. Brown. Newbury Park NJ: Sage Publications.

———. 1996. *Rhetorics, Poetics, and Cultures.* Urbana IL: NCTE.

Bevan, N. 1998. Usability Issues in Website Design. In *Proceedings of Usability Professionals' Association (UPA).* Bloomington IL: UPA.

Birkerts, S. 1994. *The Gutenberg Elegies: The Fate of Reading in an Electronic Age.* New York: Fawcett Columbine.

Blakeslee, A. 2001. Bridging the Workplace and the Academy: Teaching Professional Genres Through Classroom-Workplace Collaboration. *Technical Communication Quarterly* 10:169–92.

Blomberg, J., S. Suchman, and R. Trigg. 1996. A Work-Oriented Design Project. *Human-Computer Interaction* 11:237–65.

Bloomberg News. 2001. Sharp to Challenge Palm, Pocket PC with Linux PDA. http://news.cnet.com/news/0-1006-200-5023907.html.

Blyler, N. 1993. Theory and Curriculum: Reexamining the Curricular Separation of Business and Technical Communication. *Journal of Business and Technical Communication* 7:218–45.

Blyler, N., and C. Thralls. 1993. *Professional Communication: The Social Perspective.* Newbury Park CA: Sage.

Blythe, S. 2001. *Designing Online Courses: User-Centered Practices. Computers and Composition* 16:329–46.

Boal, A. 1968. *Theatre of the Oppressed.* Trans. C. A. McBride and M. L. McBride. New York: Theatre Communications Group.

———. 1997. *Games for Actors and Non-Actors.* Trans. A. Jackson. London: Routledge.

Boiarsky, C., and M. Dobberstein. 1998. Teaching Documentation Writing: What Else Students—and Instructors—Should Know. *Technical Communication* 45:38–46.

Bolter, J. D., and R. Grusin. 1999. *Remediation.* Cambridge: MIT Press.

Bosley, D. 1995. Collaborative Partnerships: Academia and Industry Working Together. *Technical Communication* 4:611–19.

Boyer, E. 1981. *Higher Learning in the Nation's Service.* Washington, D.C.: Carnegie Foundation for the Advancement of Teaching.

———. 1990. *Scholarship Reconsidered.* Princeton: Carnegie Foundation for the Advancement of Teaching.

———. 1997. *Scholarship Reconsidered: Priorities of the Professoriate.* Pittsburgh: Carnegie Foundation.

Boyte, H. 1993. What is Citizenship Education? *Rethinking Tradition: Integrating Service with Academic Study on College Campuses.* Denver: Education Commission of the States.

Brake, T., D. M. Walker, and T. Walker. 1995. *Doing Business Internationally: The Guide to Cross-Cultural Success.* New York: Irwin.

Brandt, D. 1995. Accumulating Literacy: Writing and Learning to Write in the Twentieth Century. *College English* 57:649–68.

Brecht, B. 1968. Kleines Organon f¸r das Theater. *Gesammelte Werke Schriften zum Theater* vol.16:2. Frankfurt A. M.: Suhrkamp

Bridgeford, T. 2002. *Narrative Ways of Knowing: Re-imagining Technical Communication Instruction.* Ph.D. diss., Michigan Technological University.

Bringle, R. G., R. Games, and E. A. Malloy. 1999. *Colleges and Universities as Citizens: Reflections.* Boston: Allyn and Bacon.

Bringle, R. G., and J. A. Hatcher. 1996. Implementing Service Learning in Higher Education. *Journal of Higher Education* 67:221–39.

Britton, J., et al. 1975. *The Development of Writing Abilities* (11–18). London: Macmillan.

Brooks, R. M. 1995. Technical Communication and Service Learning: Integrating Profession and Community. In *1995 Proceedings of the Council of Programs in Technical and Scientific Communication,* ed. M. M. Cooper. Houghton MI: Council for Programs in Technical and Scientific Communication.

Bruce, B., and A. Rubin. 1993. *Electronic Quills: A Situated Evaluation of Using Computers for Writing Classrooms.* Hillsdale NJ: Lawrence Erlbaum.

Bruffee, K. 1986. Social Construction, Language, and the Authority of Knowledge: A Bibliographic Essay. *College English* 48:773–90.

Bruner, J. 1991. The Narrative Construction of Reality. *Critical Inquiry* 18:1–21.

———. 1990. *Acts of Meaning.* Cambridge: Harvard University Press.

Burnett, R. E. 1993. Conflict in Collaborative Decision-Making. In *Professional Communication: The Social Perspective,* ed. N. Blyler and C. Thralls. Newbury Park CA: Sage.

———. 1994. *Technical Communication.* Belmont CA: Wadsworth.

———. 1997. *Technical Communication.* 4th ed. Belmont CA: Wadsworth.

Bush-Bacelis, J. L. 1998. Innovative Pedagogy: Academic Service-Learning for Business Communication. *Business Communication Quarterly* 61:20–34.

Campbell, K. 1999. Collecting Information: Qualitative Research Methods for Solving Workplace Problems. *Technical Communication* 46:532–45.

Cates-Melver, L. 1999. Internships and Co-op Programs: A Valuable Combination for Collegians. *Black Collegian* 20:85–87.

Cha, S., and M. Rothman. 1994. *Service Matters.* Providence: Brown University

Charney, D. H., and R. Rayman. 1989. The Role of Writing Quality in Effective Student resumes. *Journal of Business and Technical Communication* 3:36–53.

Chickering, A. W., and S. C. Ehrmann. 1998. Implementing the Seven Principles: Technology as Lever. *American Association for Higher Education.* Available online at http://www.aahe.org/tech-nology/ehrmann.html.

Chickering, A. W., and Z. F. Gamson. 1987. Seven Principles for Good Practice in Undergraduate Education. *AAHE Bulletin* 39:3–7.

Cilenger, E. N. 1992. Controlling Technology Through Communication: Redefining the Role of the Technical Communicator. *Technical Communication* 39:166–74.

Clark, G. 1990. *Dialogue, Dialectic, and Conversation: A Social Perspective on the Function of Writing.* Carbondale IL: Southern Illinois University Press.

Clemens, L. 1999. Preparing *Technical Communication Ph.D. Students to Teach at a Distance: Guidelines and Principles.* Master's thesis, University of Minnesota.

Cohen, A., and F. Brawer. 1989. *The American Community College.* 2nd ed. San Francisco: Jossey-Bass.

Cohen, J., and D. Kinsey. 1994. Doing Good and Scholarship: A Service-Learning Study. *Journalism and Mass Communication Educator* (winter: 4–14).

Coney, M. B. 1997. Technical Communication Theory: An Overview. In *Foundations for Teaching Technical Communication: Theory, Practice, and Program Design*, ed. K. Staples and C. Ornatowski. Greenwich CT: Ablex.

Cooper, A. 1999. *The Inmates are Running the Asylum: Why High Tech Products Drive Us Crazy and How To Restore the Sanity.* Indianapolis: SAMS.

Cooper, M. M. 1996. The Postmodern Space of Operator's Manuals. *Technical Communication Quarterly* 5, no. 4 (fall): 385–410.

Coppola, J. F., and B. A. Thomas. 2000. A Model for E-Classroom Design Beyond "Chalk And Talk." *T.H.E. Journal* 27:30–36.

Couture, B. 1998. *Toward a Phenomenological Rhetoric: Writing, Profession, and Altruism.* Carbondale. Ill.:Southern Illinois University Press.

Crawford, K. 1993. Community Service Writing in an Advanced Composition Class. In *Praxis I: A Faculty Casebook on Community Service Learning*, ed. J. Howard. Ann Arbor: OCSL Press.

Cushman, E. 1996. The Rhetorician as Agent of Social Change. *College Composition and Communication* 47:7–28.

David, C., and D. Kienzler. 1999. Towards an Emancipatory Pedagogy in Service Courses and User Departments. *Technical Communication Quarterly* 8:269–84.

Deal, T., and A. Kennedy. 1982. *Corporate Culture: The Rites and Rituals of Corporate Life.* Menlo Park CA: Addison-Wesley.

De Certeau, M. 1984. *The Practice of Everyday Life.* Trans. S. Rendall. Berkeley: University of California Press.

Denning, P. 1992. Educating a New Engineer. *Communications of the ACM* 35:83–97.

Dicks, R.S. 1999. Technical Communication in Academia and Industry: The Cultural Gaps That Prevent Understanding. Paper presented at the second annual meeting of the Association of Teachers of Technical Writing, Atlanta, Ga.

Dicks, R. S., and B. Mehlenbacher. 1999. Usability Testing "Ask NC State." *Summary Report for North Carolina State Extension, Research, and Outreach Project, October 15–June 30.* Raleigh, N.C.: North Carolina State, October 15-June 30.

Dobrin, D. N. 1989. *Writing and Technique.* Urbana IL: NCTE.

Donnell, J. A., J. Petraglia-Bahri, and A. C. Gable. 1999. Writing Vs. Content, Skills Vs. Rhetoric: More and Less False Dichotomies. *Journal of Language and Learning across the Disciplines* 3:113–17.

Dragga, S. 2001, summer. Ethics in Technical Communication, special issue. Vol. 10, #3

Dubinsky, J. 1998. *Learning the Möbius Loop of Theory and Practice: Reflections on the Techné of Teaching Writing.* Ph.D. diss., University Of Ohio. Oxford, Ohio.

———. 2002. Service-Learning as a Path to Virtue: The Ideal Orator in Professional Communication. *Michigan Journal of Community Service Learning* 8.2(2002):61–74.

Duin, A. H., L. Baer., and D. Starke-Meyerring. 2001. *Partnering in the Learning Marketspace.* San Francisco: Jossey-Bass.

Durst, R. 1999. *Collision Course.* Urbana IL: NCTE.

Eastman, D. V. 1998. Adult Learners and Internet-Based Distance Education. In *Adult Learning and the Internet*, ed. B. Cahoon. San Francisco: Jossey-Bass.

Ecker, P. S., and K. Staples. 1997. Collaborative Conflict and the Future: Academic-Industrial Alliances and Adaptations. In *Nonacademic Writing: Social Theory and Technology*, ed. A. H. Duin and C. Hansen. Mahwah NJ: Lawrence Erlbaum.

Ede, L. 1984. Audience: An Introduction to Research. *College Composition and Communication* 35:140–54.

Ede, L., and A. Lunsford. 1984. Audience Addressed/Audience Invoked: The Role of Audience in Composition Theory and Pedagogy. *College Composition and Communication* 35:155–71.

Ellsworth, N. J., C. N. Hedley, and A. N. Barbatta, eds. 1994. *Literacy: A Redefinition.* Hillsdale NJ: Lawrence Erlbaum.

Emig, J. 1971. *The Writing Process of Twelfth Graders.* Urbana IL: NCTE.

English, D., and D. Koeppen. 1993. The Relationship of Accounting Internships and Subsequent Academic Performance. *Issues in Accounting Education* 8:292–300.

Faber, B. 1999. Intuitive Ethics: Understanding and Critiquing the Role of Intuition in Ethical Decisions. *Technical Communication Quarterly* 8:189–203.

Faigley, L., R. Cherry, D. Joliffe, and A. M. Skinner. 1985. *Assessing Writers' Knowledge and Processes of Composing.* Norwood NJ: Ablex.

Fearing, B. E. and W. K. Sparrow. 1989. *Technical Writing: Theory and Practice.* New York: MLA.

Flower, L. 1997. Partners in Inquiry: A Logic for Community Outreach. In *Writing the Community: Concepts and Models for Service-Learning in Composition*, ed. L. Adler-Kasner, R. Crooks, and A. Watters. Washington DC: AAHE/NCTE.

Flynn, E. 1997. Emergent Feminist Technical Communication. *Technical Communication Quarterly* 6:313–20.

Flynn, E. A., R. W. Jones, D. Shoos, and B. Barna. 1990. Michigan Technological University. In *Programs that Work: Models and Methods for Writing across the Curriculum*, ed. T. Fulwiler and A. Young. Portsmouth NH: Boynton/Cook Heinemann.

Forman, J. 1993. Business Communication and Composition: The Writing Connection and Beyond. *Journal of Business Communication* 30:333–52.

Foucault, M. 1977a. The Order of Discourse. Trans. I. McLeod. In *The Rhetorical Tradition*, ed. P. Bizzell and B. Herzberg. Boston: Bedford Books.

———. 1977b. Truth and Power. In *The Foucault Reader*, ed. P. Rabinow. New York: Pantheon Books.

Fulop, M. P., and N. N. Varzandeh. 1996. The Role of Computer-Based Resources in Health Promotion and Disease Prevention: Implications for College Health. *Journal of American College Health* 45:11–17.

Fulwiler, T. 1991. The Quiet and Insistent Revolution: Writing across the Curriculum. In *The Politics of Writing Instruction: Postsecondary*, ed. R. Bullock and J. Trimbur. Portsmouth NH: Boynton/Cook Publishers.

Garay, M. S., and S. A. Bernhardt. 1998. *Expanding Literacies: English Teaching and the New Workplace.* New York: SUNY.

Gazzaniga, M. 1998. How to Change the University. *Science* 282:237.

Gee, J. P., G. Hull, and C. Lankshear. 1996. *The New Work Order: Behind the Language of the New Capitalism.* Boulder CO: Westview Press.

Gehrke, R. 2002. Hispanic School-Age Population Fastest-Growing, Report Says. *El Paso Times*, June 20, 4A.

George, D., and D. Shoos. 1992. Issues of Subjectivity and Resistance: Cultural Studies in the Composition Classroom. In *Cultural Studies in the English Classroom*, ed. J. A. Berlin and M. J. Vivion. Portsmouth NH: Boynton/Cook Heinemann.

Gilchrist, C. 1997. *Faculty Attitudes and Perceptions Toward Using Interactive Television: A Case Study*. Ph.D. diss., University of Minnesota, Twin Cities.

Giles, D.E., Honnet, E. Porter, and S. Migliore. 1991. *Research Agenda for Combining Service and Learning in the 1990s*. Raleigh NC: National Society for Internships and Experiential Education.

Gilsdorf, J., and D. Leonard. 2001. Big Stuff, Little Stuff: A Decennial Measurement of Executives' and Academics' Reactions to Questionable Usage Elements. *The Journal of Business Communication* 38:439–75.

Goodlad, J. I., and P. Keating, eds. 1994. *Access to Knowledge: The Continuing Agenda for Our Nations' Schools*. New York: College Entrance Examination Board.

Grabill, J. T. 2000. Shaping Local HIV/AIDS Services Policy Through Activist Research: The Problem of Client Involvement. *Technical Communication Quarterly* 9:29–50.

Grice, R. 1987. *Technical Communication in the Computer Industry: An Information- Development Process to Track, Measure, and Ensure Quality*. Ph.D. diss., Rensselaer Polytechnic Institute.

Halloran, S. M. 1978. Technical Communication and the Rhetoric of Science. *Journal of Technical Writing and Communication* 8:77–88.

Hanna, D. E. and Associates. 2000. *Higher Education in an Era of Digital Competition: Choices and Challenges*. Madison, Wisc.: Atwood Publishing.

Hansen, C. 1995. Writing the Project Team: Authority and Intertextuality in a Corporate Setting. *Journal of Business Communication* 32:103–23.

Harris, E. 1980. Response to Elizabeth Tebeaux. *College English* 41:827–29.

———. 1982. In Defense of the Liberal-Arts Approach to Technical Writing. *College English* 44, no. 6:628–36.

Hartung, K. K. 1998. What Are Students Being Taught about the Ethics of Technical Communication?: An Analysis of the Ethical Discussions Presented in Four Textbooks. *Journal of Technical Writing and Communication* 28:363–83.

Haussamen, B. 1997. Service-Learning and First-year Composition. *Teaching in the Two Year College*. 24, no. 3:192–98.

Hawisher, G., and C. L. Selfe, eds. 1999. *Passions, Pedagogies, and Twenty-First Century Technologies*. Urbana: NCTE.

Hayhoe, G. F. 1998. The Academe-Industry Partnership: What's in It for All of Us? *Technical Communication* 45, no. 1:19–20.

Head, A. J. 1999. *Design Wise: A Guide for Evaluating the Interface Design of Information Resources*. Medford NJ: Cyberage Books.

Henson, L., and K. Sutliff. 1998. A Service Learning Approach to Business and Technical Writing Instruction. *Journal of Technical Writing and Communication* 28:189–205.

Herndl, C. 1991. Writing Ethnography: Representation, Rhetoric, and Institutional Practices. *College English* 53:320–32.

———. 1993a. Cultural Studies and Critical Science. In *Understanding Scientific Prose*, ed. J. Selzer. Madison WI: University of Wisconsin Press.

———. 1993b. Teaching Discourse and Reproducing Culture: A Critique of Research and Pedagogy in Professional and Non-Academic Writing. *College Composition and Communication* 44:349–63.

———. 1996a. Tactics and the Quotidian: Resistance and Professional Discourse. *Journal of Advanced Composition* 16:455–70.

———. 1996b. The Transformation of Critical Ethnography into Pedagogy, or the Vicissitudes of Traveling Theory. In *Nonacademic Writing: Social Theory and Technology*, ed. A. Duin and C. Hansen. Mahweh NJ: Lawrence Erlbaum.

Herzberg, B. 1994. Community Service and Critical Teaching. *College Composition and Communication* 45:307–19.

Hogan, Harriet. 1983. Distinguishing Characteristics of the Technical Writing Course. In *Technical and Business Communication in Two-Year Programs*, ed. K. W. Sparrow and N. A. Pickett. Urbana IL: NCTE.

Holland, V. M., V. R. Charrow, and W. W. Wright. 1988. How Can Technical Writers Write Effectively for Several Audiences at Once? In *Solving Problems in Technical Writing*, ed. L. Beene and P. White. New York: Oxford University Press.

Honebein, P., T. M. Duffy, and B. Fishman. 1993. Constructivism and the Design of Learning Environments: Context and Authentic Activities for Learning. In *Designing Environments for Constructivist Learning*, ed. T. M. Duffy, J. Lowyck, and D. Jonassen. Heidelberg: Springer-Verlag.

Huckin, T. 1997. Technical Writing and Community Service. *Journal of Business and Technical Communication* 11:49–60.

Hudson, L. P. 1984. *The Bones of Plenty*. St. Paul MN: Minnesota Historical Society Press.

Huizinga, J. 1990. The Nature of Play. In *Philosophic Inquiry in Sport*, pp. 3-6.

International Society for Technology in Education. 2000. *National Education Technology Standards for Students: Connecting Curriculum and Technology*. Eugene, Ore.

Johnson, D. W., and R. Johnson. 1998. *Cooperation and Competition: Theory and Research*. Englewood Cliffs NJ: Prentice Hall.

Johnson, R. R. 1998a. Complicating Technology: Interdisciplinary Method, the Burden of Comprehension, and the Ethical Space of the Technical Communicator. *Technical Communication Quarterly* 7:75–98.

———. 1998b. *User-Centered Technology: A Rhetorical Theory for Computers and Other Mundane Artifacts*. Albany: SUNY Press.

———. 1999. Johnson Responds. *Technical Communication Quarterly* 8:223–26.

Johnson-Eilola, J. 1996. Relocating the Value of Work: Technical Communication in a Post-Industrial Age. *Technical Communication Quarterly* 5:245–71.

———. 1997. Wild Technologies: Computer Use and Social Possibility. In *Computers and Technical Communication: Pedagogical and Programmatic Perspectives*, ed. S. A. Selber. Greenwich CT: Ablex.

Jorn, L., A. H. Duin, and B. J. Wahlstrom. 1996. Designing and Managing Virtual Learning Communities. *IEEE Transactions on Professional Communication* 39:183–91.

Kaasbøll, J. J. 1998. Teaching Critical Thinking and Problem Defining Skills. *Education and Information Technologies* 3:101–17.

Kahne, J., and J. Westheimer. 1996. In Service of What? The Politics of Service Learning. *Phi Delta Kappan* 77:593–600.

Kalmbach, J. 1997. From Liquid Paper to the Typewriter: Some Historical Perspectives on Technology in the Classroom. *Computers and Composition* 13:57–68.

Kantrowitz, B., and A. Rogers. 1994. The Birth of the Internet. *Newsweek*, August 8, 56–58.

Karis, B. 1997. Building Relationships to Garner Technological Resources and Support in Technical Communication Programs. In *Computers and Technical Communication: Pedagogical and Programmatic Perspectives*, ed. S. A. Selber. Greenwich CT: Ablex.

Karis, W. M. 1989. Using Literature to Focus Attention: Rhetorical Models and Case Studies. *The Technical Writing Teacher* 16:187–94.

Karras, T. 1999. The Mammogram Screening Controversy: When Should You Start? http://www.cnn.com/HEALTH/women/9909/27/bcam.mammography/.

Kastman Breuch, L. 2001. The Overruled Dust Mite: Preparing Students to Interact with Clients. *Technical Communication Quarterly* 10:193–210.

Katz, S. 1992. The Ethics of Expediency: Classical Rhetoric, Technology, and the Holocaust. *College English* 54:255–75.

Keene, M. L. 1997. *Education in Scientific and Technical Communication: Academic Programs that Work.* Arlington, Va.: Society for Technical Communication.

Kendall, J. C. 1990. Combining Service and Learning: An Introduction. In *Combining Service and Learning: A Resource Book for Community and Public Service,* ed. J. C. Kendall and Associates. Raleigh NC: NSEE.

Kienzler, D. 2001. Ethics, Critical Thinking, and Professional Communication Pedagogy. *Technical Communication Quarterly* 10:319–40.

Kilgore, D. 1981. Moby-Dick: A Whale of a Handbook for Technical Writing Teachers. *Journal of Technical Writing and Communication* 11:209–16.

Killingsworth, M. J. 1997. Developing Programs in Technical Communication: A Pragmatic View. In *Foundations for Technical Communication: Theory, Practice, and Program Design,* ed. K. Staples and C. Ornatowski. Greenwich CT: Ablex.

Killingsworth, M. J., and J. S. Palmer. 1999. *Information in Action.* 2nd ed. Boston: Allyn and Bacon.

Kim, L. And M. J. Albers. 2002, August. "Web Design Issues When Searching for Information Using Handheld Interfaces". *Technical Communication.* Vol. 49, no. 3.

Kitalong, K. S., D. Selfe, and M. Moore. Forthcoming. Technology Autobiographies and Student Participation in English Studies Literacy Classes. *Teaching/Writing in the Late Age of Print,* ed. J. R. Galin, C. P. Haviland, and J. P. Johnson. Kresskill NJ: Hampton Press.

Knouse, S., J. Tanner, and E. Harris. 1999. The Relation of College Internships, College Performance, and Subsequent Job Opportunity. *Journal of Employment Counseling* 36:35–44.

Knox, E. L. 1997. The Pedagogy of Web Site Design. *ALN Magazine* vol.1, no. 2. http://www.aln.org/alnWeb/magazine/issue2/know.htm.

Kolb, D. A. 1984. *Experiential Learning: Experience as the Source of Learning and Development.* Englewood Cliffs NJ: Prentice Hall.

Koschmann, T., A. C. Kelson, P. J. Feltovich, and H. S. Barrows. 1996. Computer-Supported Problem-Based Learning: A Principled Approach to the Use of Computers in Collaborative Learning. In *CSCL: Theory and Practice of an Emerging Paradigm,* ed. T. Koschmann. Mahwah NJ: Lawrence Erlbaum.

Koski, C. A. 1997. Down the Rabbit-Hole: Exploring Health Messages on the World Wide Web. *Journal of Technical Writing and Communication* 27:49–55.

Krause, T. 1997. Preparing an Online resume. *Business Communication Quarterly* 60:159–61.

Kretzmann, J. P., and J. L. McKnight. 1993. *Building Communities from the Inside Out: A Path Toward Finding and Mobilizing a Community's Assets.* Chicago: ACTA Publications.

Kryder, L. 1999. Mentors, Models, and Clients: Using the Professional Engineering Community to Identify and Teach Engineering Genres. *IEEE Transactions on Professional Communication* 42:3–12.

Kunin, M. 1997. Service Learning and Improved Academic Achievement. *Service Learning*, ed. J. Schine. Chicago: NSSE.

Kynell, T. 1996. *Writing in a Milieu of Utility*. Norwood NJ: Ablex.

Landauer, T. K. 1995. *The Trouble with Computers: Usefulness, Usability, and Productivity*. Cambridge: MIT Press.

Lanham, R. 1983. One, Two, Three. *Composition and Literature*, ed. W. Horner. Chicago: University of Chicago Press.

Latour, B. 1993. *We Have Never Been Modern*. Trans. C. Porter. Cambridge: Harvard University Press.

Laurillard, D. 2002. Rethinking Teaching for the Knowledge Society. *Educause* 37:16–25.

Lay, M. M., B. J. Wahlstrom, S. Doheny-Farina, A. H. Duin, S. B. Little, C. D. Rude, C. L. Selfe, and J. Seltzer. 1995. *Technical Communication*. Chicago: Irwin.

Lee, Y. S. 1998. University-Industry Collaboration on Technology Transfer: Views from the Ivory Tower. *Policy Studies Journal* 26:68.

Leigh, J. W. 1998. *Communicating for Cultural Competence*. Boston: Allyn and Bacon.

Lerner, I. J., and B. J. Kennedy. 1992. The Prevalence of Questionable Methods of Cancer Treatment in the United States. *CA-A Cancer Journal* 42:181–91.

Levy, S. 1984. *Hackers*. New York: Dell.

Longo, B. 1998. An Approach for Applying Cultural Study Theory to Technical Writing Research. *Technical Communication Quarterly* 7:53–73.

Lutz, J., and M. Fuller. 1998. The Cure From Within: The Rhetoric of Alternative Medicines. Paper presented at the Conference on College Composition and Communication, April.

Lynch, D. A. 1997. Email in an Interdisciplinary Context. In *Electronic Communication across the Curriculum*, ed. D. Reiss, D. Selfe, and A. Young. Urbana IL: NCTE.

Mansfield, M. A. 1993. Real World Writing and the English Curriculum. *College Composition and Communication* 44:69–83.

Marchionini, G., and C. Hert. 1997. Usability Testing Large Institutional Websites. *Usability Testing World Wide Web Sites: Position Papers From a Two-Day Workshop at CHI 97*. March 23–24 in Atlanta, Ga.. Available online at http://www.acm.org/sigchi/webhci/chi97/testing/marchion.htm.

Markel, M. 1997. Ethics and Technical Communication: A Case for Foundational Approaches. *IEEE Transactions on Professional Communication* 40:84–99.

———. 1999. *Journal of Business and Technical Communication* 13:208–22.

Markus, G. B., J. P. F. Howard, and D. C. King. 1993. Integrating Community Service and Classroom Instruction Enhances Learning: Results from an Experiment. *Educational Evaluation and Policy Analysis* 15:410–19.

Martin, W. B. 1977. Teaching, Research, and Service—But the Greatest of These is Service. *Redefining Service, Research, and Teaching*, ed. W. B. Martin. San Francisco: Jossey- Bass.

Matthews, C., and B. B. Zimmerman. 1999. Integrating Service Learning and Technical Communication: Benefits and Challenges. *Technical Communication Quarterly* 8:383-404.

Mauriello, N., G. S. Pagnucci, and T. Winner. 1999. Reading between the Code: The Teaching of HTML and the Displacement of Writing Instruction. *Computers and Composition* 16:409–19.

Mawby, R. G. 1996. The Challenge for Outreach for Land-Grant Universities as They Move into the Twenty-First Century. *Journal of Public Service and Outreach* 1:46–56.

McCafferty, D. 2002. Dude, What's in Your Car? *USA Weekend* 26:6–7.

McCormack, C., and D. Jones. 1998. *Building a Web-Based Education System.* New York: Wiley.

McCune, J.C. 2000. Training Drain. *Management Review,* March 30. Available online at http://web7.infotrac.galegroup.com.

McDowell, E. E. 1987. Perceptions of the Ideal Cover Letter and Ideal Resume. *Journal of Technical Writing and Communication* 17:179–91.

McEachern, R. 2001. Problems in Service Learning and Technical/Professional Writing: Incorporating the Perspective of Nonprofit Management. *Technical Communication Quarterly* 10:210–24.

McKnight, J. 1995. *The Careless Society: Community and Its Counterfeits.* New York: Basic Books.

Mead, M. 1970. *Culture and Commitment: A Study of the Generation Gap.* Garden City NY: Natural History Press/Doubleday.

Mehlenbacher, B. 1997. Technologies and Tensions: Designing Online Environments for Teaching Technical Communication. In *Computers and Technical Communication: Pedagogical and Programmatic Perspectives,* ed. S. A. Selber. Greenwich CT: Ablex.

Meister, J. C. 2000. Savvy Learners Drive Revolution in Education: The Case for Corporate Universities. *Financial Times* (London), April 24, 2000. Available online at http://web.lexis-nexis.com/universe.

———. 2001. The Brave New World of Corporate Education. *Chronicle of Higher Education,* February 9, B10–11.

Mellander, G., and N. Mellander. 1998. Corporate America: Inroads Realizing the Dream; Corporate Internships Benefit Thousands. *Hispanic Outlook in Higher Education* 8:19.

Meyer, P. R., and S. A. Bernhardt. 1998. Workplace Realities and the Technical Communication Curriculum: A Call for Change. In *Foundations for Teaching Technical Communication: Theory, Practice, and Program Design,* ed. K. Staples and C. Ornatowski. Greenwich CT: Ablex.

Michaels, A. 2000. Companies Get Hit by the Learning Bug: The Growth in the Number of Corporate Universities Reflects a Sea of Change in the Training of Executives and Employees. *The Financial Times* (London), April 24, 2000. Available online at http://web.lexis-nexis.com/universe.

Miller, C. R. 1979. A Humanistic Rationale for Technical Writing. *College English* 40:610–17.

———. 1984. Genre as Social Action. *Quarterly Journal of Speech* 70:151–67.

———. 1989. What's Practical about Technical Writing. In *Technical Writing: Theory and Practice,* ed. B. E. Fearing and W. K. Sparrow. New York: MLA.

———. 1996. Comments on "Instrumental Discourse Is As Humanistic As Rhetoric." *Journal of Business and Technical Communication* 10:482–86.

Miller, G. 1996, Oct. 7. Gap exists between net awareness and use. *Los Angeles Times,* online at http://www.latimes.com

MIRA (Managing Information with Rural America). 2002. Project Scrapbook, May 31. Available online at http://mira.wkkf.org/about.htm.

Mirel, B. 1993. Beyond the Monkey House: Audience Analyses in Computerized Workplaces. In *Writing in the Workplace: New Research Perspectives,* ed. R. Spilka. Carbondale and Edwardsville, IL: Southern Illinois University Press.

———. 1998. "Applied Constructivism" for User Documentation: Alternatives to Conventional Task Orientation. *Journal of Business and Technical Communication* 12:7–49.

Mirel, B., and R. Spilka. 2002. *Reshaping Technical Communication: New Directions and Challenges for the Twenty-First Century.* Mahwah NJ: Lawrence Erlbaum.

MLA Commission on Professional Service. 1996. Making Faculty Work Visible: Reinterpreting Professional Service, Teaching, and Research in the Fields of Language and Literature. *Profession*, 161–216.

Moffett, J. 1968. *Teaching the Universe of Discourse*. Boston: Houghton.

Moore, P. 1996. Instrumental Discourse Is As Humanistic As Rhetoric. *Journal of Business and Technical Communication* 10:100–118.

———. 1999. Myths about Instrumental Discourse: A Response to Robert R. Johnson. *Technical Communication Quarterly* 8:210–26.

Moran, M. H., and M. G. Moran. 1985. Business Letters, Memoranda, and Resumes. In *Research in Technical Communication*, ed. M. Moran and D. Journet. Westport, CT: Greenwood.

Moulthrop, S. 1999. Everybody's Elegies. In *Passions, Pedagogies and Twenty-First Century Technologies*, ed. G. Hawisher and C. Selfe, Urbana IL: NCTE.

Mowbray, N. 2000. *The Rhetoric of Acupuncture*. Master's thesis, James Madison University, Harrisonburg, Virginia.

Murray, D. 1972. Teach Writing as a Process Not Product. *The Leaflet*: 11–14.

Myers, G. 1990. The Double Helix as Icon. *Science as Culture* 9:49–72.

NCTE and the International Reading Association. 1996. *Standards for the English Language Arts*. Urbana IL: NCTE

Nagelhout, E. 1999. Pre-Professional Practices in the Technical Writing Classroom: Promoting Multiple Literacies through Research. *Technical Communication Quarterly* 8:285–300.

National Cancer Institute. 2000. Available online at http://cancernet.nci.nih.gov.

Neff, J. M. 1998. From a Distance: Teaching Writing on Interactive Television. *Research in the Teaching of English* 3:136–57.

Nelkin, D. 1995. The Press on the Technological Frontier. In *Selling Science: How the Press Covers Science and Technology*. New York: W. H. Freeman.

Nemnich, M. B., and F. E. Jandt. 1999. *Cyberspace Resume Kit*. Indianapolis: JIST.

Newman, F. 1985. *Higher Education and the American Resurgence*. Princeton NJ: Carnegie Foundation for the Advancement of Teaching.

Nielsen, J. 1994. Heuristic Evaluation. In *Usability Inspection Methods*, ed. J. Nielsen and R. L. Mack. New York: John Wiley and Sons.

———. 1997. Usability Engineering. In *The Computer Science and Engineering Handbook*, ed. A. B. Tucker, Jr. Boca Raton FL: CRC Press.

Noble, D. F. 1997. *Digital Diploma Mills: The Automation of Higher Education*. New York: Monthly Review Press.

Norman, D. 1990. *The Design of Everyday Things*. New York: Doubleday.

———. 1994. *Things That Make Us Smart: Defending Human Attributes in the Age of the Machine*. Cambridge MA: Perseus Publishing.

Øgrim, L. 1991. Project Work in System Development Education. In *Information System, Work, and Organization Design*, ed. P. van den Besselaar, A. Clement, and P. Jårvinen. Amsterdam: North-Holland.

Ormiston, G. L., ed. 1990. *From Artifact to Habitat: Studies in the Critical Engagement of Technology*. Bethlehem PA: Lehigh University Press.

Ornatowski, C. M. 1997. Technical Communication and Rhetoric. In *Foundations for Teaching Technical Communication*, ed. K. Staples and C. Ornatowski. Greenwich CT: Ablex.

Parker-Gwin, R., and J. B. Mabry. 1998. Service Learning as Pedagogy and Civic Education: Comparing Outcomes for Three Models. *Teaching Sociology* 26:276–91.

Perelman, S. J. 1976. Insert Flap "A" and Throw Away. In *Humor in America: An Anthology*, ed. E. Veron. San Diego: Harcourt Brace Jovanovich.

Pew Higher Education Roundtable. 1994. To Dance with Change. *Policy Perspectives* 5:1A–12A.

Phelps, L. W. 1991. Practical Wisdom and the Geography of Knowledge in Composition. *College English* 53:863–85.

Phillips, G. W., and L. Metzler. 1991. The Corporate-Academic Relationship: Risks and Returns. *Fund Raising Management* 22:26–30.

Pirsig, R. M. 1974. *Zen and the Art of Motorcycle Maintenance.* New York: Bantam.

Plauche, C. 2000. The J.M. Smucker Company Named to Fortune's List of "100 Best Companies to Work For" For Third Year in a Row. Available online at www.smucker.com/news_fortune.html.

Popken, R. 1999. The Pedagogical Dissemination of a Genre: The resume in American Business Discourse Textbooks, 1914–1939. *Journal of Advanced Composition* 19:91–116.

Porter, J. E., and P. Sullivan. 1996. Working across Methodological Interfaces: The Study of Computers and Writing in the Workplace. In *Electronic Literacies in the Workplace: Technologies of Writing*, ed. P. Sullivan and J. Dautermann. Urbana IL and Houghton MI: NCTE/Computers and Composition.

Porter, L. 1997. *Creating the Virtual Classroom: Distance Learning with the Internet.* New York: Wiley.

Powers, D. R., M. F. Powers, F. Betz, and C. B. Aslanian. 1998. *Higher Education in Partnership with Industry: Opportunities and Strategies for Training, Research, and Economic Development.* San Francisco: Jossey-Bass.

Press, E., and J. Washburn. 2000. The Kept University. *The Atlantic Monthly* 285:39–54.

Quibble, Z. K. 1995. Electronic resumes: Their Time Is Coming. *Business Communication Quarterly* 58:5–9.

Redish, J. C. 1988. Reading to Learn to Do. *The Technical Writing Teacher* 15, no. 3(fall): 223–33. Reprinted in *IEEE Transactions on Professional Communication* 32, no. 4(December): 289–93.

———. 1993. Understanding Readers. In *Techniques for Technical Communicators*, ed. C. M. Barnum and S. Carliner. Needham Heights MA: Allyn and Bacon.

———. 1997. Understanding People: The Relevance of Cognitive Psychology to Technical Communication. In *Foundations for Teaching Technical Communication: Theory, Practice, and Program Design*, ed. K. Staples and C. Ornatowski. Greenwich CT: Ablex.

Rehling, L. 1998. Exchanging Expertise: Learning from the Workplace and Educating It Too. *Journal of Writing and Technical Communication* 28:385–93.

Rendon, L. I. 1994. Validating Culturally Diverse Students: Toward a New Model of Learning and Student Development. *Innovative Higher Education* 19:33–51.

Reynolds, J. F., C. B. Matalene, J. N. Magnotto, D. C. Samson, and L.V. Sadler. 1995. *Professional Writing in Context: Lessons from Teaching and Consulting in Worlds of Work.* Hillsdale NJ: Lawrence Erlbaum.

Rogoff, B. 1990. *Apprenticeship in Thinking: Cognitive Development in Social Context.* NY, NY: Oxford University Press.

Ronald, K. 1987. The Politics of Teaching Professional Writing. *Journal of Advanced Composition* 7:23–30.

Rooney, A. 1997. Warning: Do Not Put This Column in Water. *Daily Mining Gazette*, July 26, 4A.

Rubin, J. 1994. *Handbook of Usability Testing: How to Plan, Design, and Conduct Effective Tests.* New York: John Wiley and Sons.

Russell, D. R. 1991. *Writing in the Academic Disciplines, 1870–1990.* Carbondale and Edwardsville IL: Southern Illinois University Press.

Samovar, L. S., and R. E. Porter. 1988. *Intercultural Communication: A Reader.* Belmont CA: Wadsworth.

Samuelson, R. J. 2002. Debunking the Digital Divide. *Newsweek*, March 25, 37.

Sanders, S. R. 1985. *Terrarium.* Bloomington and Indianapolis: Indiana University Press.

Savery, J. R. 1998. Fostering Ownership for Learning With Computer-Supported Collaborative Writing in an Undergraduate Business Communication Course. In *Electronic Collaborators: Learner-Centered Technologies for Literacy, Apprenticeship, and Discourse*, ed. C. J. Bonk and K. S. King. Mahwah NJ: Lawrence Erlbaum.

Sawicki, D. S., and W. Craig. 1996. The Democratization of Data: Community Groups and Information Technology in the Next Decade. *Journal of the American Planning Association* 62:512–23.

Sax, L. J. 1997. Health Trends Among College Freshmen. *Journal of American College Health* 45:252–62.

Schank, R. C. 2000. A Vision for Education in the Twenty-First Century. *T.H.E. Journal* 27:42–45.

Schmidt, W. H., and J. P. Finnigan. 1992. *The Race without a Finish Line: America's Quest for Total Quality.* San Francisco: Jossey-Bass.

Schmuck, R.A., and P. A. Schmuck. 1997. *Group Processes in the Classroom.* 7th ed. Boston: McGraw-Hill.

Schön, D. A. 1983. *The Reflective Practitioner: How Professionals Think in Action.* New York: Basic Books.

———. 1987. *Educating the Reflective Practitioner: Toward a New Design for Teaching and Learning in the Professions.* San Francisco: Jossey-Bass.

Schriver, K. A. 1996. *Dynamics in Document Design: Creating Texts for Readers.* New York: John Wiley and Sons.

Schutz, A., and A. R. Gere. 1998. Service Learning and English Studies. *College English* 60:129–49.

Selber, S. A., ed. 1997. *Computers and Technical Communication: Pedagogical and Programmatic Perspectives.* Greenwich CT: Ablex.

Selber, S. A., J. Johnson-Eilola, and B. Mehlenbacher. 1997. Online Support Systems: Tutorials, Documentation, and Help. In *The Computer Science and Engineering Handbook*, ed. A. B. Tucker, Jr. Boca Raton, Fla.: CRC Press.

Selfe, C. L. 1999. Technology and Literacy: A Story About the Perils of Not Paying Attention. *College Composition and Communication* 50:411–36.

Selfe, C. L., and G. Hawisher. 2000. *Studying the Acquisition and Development of Technological Literacy: Research Report.* The Society for Technical Communication.

Selfe, R. 1998. *Critical, Technical Literacy Practices in and around Technology-Rich Communication Facilities.* Ph.D. diss., Michigan Technological University.

Selzer, J. 1983. The Composing Processes of an Engineer. *College Composition and Communication* 34:178–87.

Sherer, H. M. 1984. *Effective Entry Level Organizational Communication as Assessed Through a Survey of Personnel Recruiters.* Ph.D. diss., Indiana University.

Shirk, H. N. 1997. New Roles for Technical Communicators in the Computer Age. In *Computers and Technical Communication: Pedagogical and Programmatic Perspectives*, ed. S. A. Selber. Greenwich CT: Ablex.

Shneiderman, B. 1998. *Designing the User Interface: Strategies for Effective Human-Computer Interaction*. 3rd ed. Reading MA: Addison-Wesley Longman.

Sigmon, R. 1994. *Linking Service with Learning in Liberal Arts Education*. Washington DC: Council of Independent Colleges.

Slavin, R. 1990. *Cooperative Learning: Theory, Research, and Practice*. Englewood Hills NJ: Prentice Hall.

Soloway, E., M. Guzdial and K. Hay. 1994. Learner-Centered Design: The Challenge for the Twenty-First Century. *Interactions*: 4 no. 2:36–48.

Sosnoski, J. J. 1994. *Token Professionals and Master Critics: A Critique of Orthodoxy in Literary Studies*. Albany: State University of New York Press.

Spiro, R. J., P. J. Feltovich, R. L. Coulson, and D. K Anderson. 1989. Multiple Analogies for Complex Concepts: Antidotes for Analogy-Induced Misconception in Advanced Knowledge Acquisition. In *Similarity and Analogical Reasoning*, ed. S. Vosniadou and A. Ortony. Cambridge: Cambridge University Press.

Spiro, R. J., W. P. Vispoel, J. G. Schmitz, A. Samarapungavan, and A.E. Boerger. 1987. Knowledge Acquisition for Application: Cognitive Flexibility and Transfer in Complex Content Domains. In *Executive Control Processes in Reading*, ed. B. K. Britton and S. M. Glynn. Hillsdale NJ: Lawrence Erlbaum.

Stanton, T. K., D. E. Giles, Jr., and N. I. Cruz. 1999. *Service-Learning: A Movement's Pioneers Reflect on its Origins, Practice, and Future*. San Francisco: Jossey-Bass.

Staples, K., and C. Ornatowski. 1997. *Foundations for Teaching Technical Communication: Theory, Practice, and Program Design*. Greenwich CT: Ablex.

Steele, C. M. 1997. A Threat in the Air: How Stereotypes Shape Intellectual Identity and Performance. *American Psychologist* 52:613–29.

Sullivan, D. 1990. Political-Ethical Implications of Defining Technical Communication as a Practice. *Journal of Advanced Composition* 10:375–86.

———. 2000. Email correspondence, May 5.

Sullivan, P., and J. Porter. 1993. Remapping Curricular Geography: Professional Writing in/and English. *Journal of Business and Technical Communication* 7:389–422.

———. 1997. *Opening Spaces: Writing Technologies and Critical Research Practices*. Greenwich CT: Ablex.

Swift, C., and R. Kent. 1999. Business School Internships: Legal Concerns. *Journal of Education for Business* 75:23–7.

Tebeaux, E. 1980. Let's Not Ruin Technical Writing, Too: A Comment on the Essays of Carolyn Miller and Elizabeth Harris. *College English* 41:822–25.

———. 1985. Redesigning Professional Writing Courses to Meet the Communication Needs of Writers in Business and Industry. *College Composition and Communication* 36:419–28.

———. 1989. The High-Tech Workplace: Implications for Technical Communication Instruction. In *Technical Writing: Theory and Practice*, ed. B. E. Fearing and W. K. Sparrow. New York: MLA.

———. 1996. Nonacademic Writing into the Twenty-First Century: Achieving and Sustaining Relevance in Research and Curricula. In *Nonacademic Writing: Social Theory and Technology*, ed. A. H. Duin and C. Hansen. Mahwah NJ: Lawrence Erlbaum.

Tiffin, J., and L. Rjasingham. 1995. *In Search of the Virtual Class: Education in an Information Society*. London: Routledge.

Tocqueville, A. de. 1974. *Democracy in America*. New York: Penguin.

Tovey, J. 1991. Using Visual Theory in the Creation of Resumes: A Bibliography. *Bulletin of the Association for Business Communication* 54:97–99.

———. 2001. Building Connections between Industry and University: Implementing an Internship Program at a Regional University. *Technical Communication Quarterly* 10:225–39.

Trace, J. 1985. Teaching Resume Writing the Functional Way. *The Bulletin of the ABC* 48:74–76.

Trigg, R., and S. Anderson. Introduction to This Special Issue on Current Perspectives on Participatory Design. *Human-Computer Interaction* 11:181–85.

Trimbur, J. 1997. Whatever Happened to the Fourth C?: Composition, Communication, and Socially Useful Knowledge. Paper presented at the Conference on College Composition and Communication, Phoenix. Available online at http: www.hu.mtu.edu/~cccc/97/trimbur/.

Turner, B., and Kearns, J. 1996. Writing and Reading History: Teaching Narrative in a Linked Writing Course. *Journal of Teaching Writing* 15:3–24.

United States. 2000. *National Environmental Policy Act*. Title 42, chapter 55. Order no. 42. U.S. Code. Sec. 4321. Available online at http://archnet.uconn.edu/archnet/topical/crm/usdocs/nepa1.htm [cited September 19, 2000].

United States. 2000. *Agriculture Adjustment Act*. Title 7. U.S. Code. Sec. 601. Available online at http://www4.law.cornell.edu/uscode/7/601.html [cited September 19, 2000].

U.S. Department of Education. 1996. *Getting America's Students Ready for the Twenty-First Century: Meeting the Technology Literacy Challenge, A Report to the Nation on Technology and Education*. Washington DC.

United States. 1998. *Science and Engineering Indicators*. GOP (NSB 98-1) National Science Board. Division of Science Resources Statistics. Http://www.nsf.gov/sbe-/srs/seind/start/htm

Varner, I. 2000. The Theoretical Foundation for Intercultural Business Communication: A Conceptual Model. *The Journal of Business Communication* 37 (January): 39–57.

Venuti, L. 1998. *The Scandals of Translation*. New York: Routledge.

Wahlstrom, B. J. 1997. Teaching and Learning Communities: Locating Literacy, Agency, and Authority in a Digital Domain. In *Computers and Technical Communication: Pedagogical and Programmatic Perspectives*, ed. S. A. Selber. Greenwich CT: Ablex.

Walsh, M. E. 1977. Teaching the Letter of Application. *College Composition and Communication* 28:74–76.

Watson, K. 1992. An Integration of Values: Teaching the Internship Course in a Liberal Arts Environment. *Communication Education* 41:429–40.

Weaver, W. 1989. A Gravestone Made of Wheat. In *A Gravestone Made of Wheat*. St. Paul MN: Graywolf Press.

Weinstein, L. A. 1993. *Moving a Battleship with Your Bare Hands: Governing a University System*. Madison WI: Magna Publications.

Wells, S. 1986. Jürgen Habermas, Communicative Competence, and the Teaching of Technical Discourse. In *Theory in the Classroom*, ed. C. Nelson. Urbana IL: University of Illinois Press.

Wenger, E. 1998. *Communities of Practice: Learning, Meaning, and Identity*. Cambridge: Cambridge University Press.

Werner, M., and D. Kaufer. 1997. Guiding Technical Communication Programs through Rapid Change: The Cycle between Technological and Curricular Change. In *Computers and Technical Communication: Pedagogical and Programmatic Perspectives*, ed. S. A. Selber. Greenwich CT: Ablex.

Whitburn, M. 1984. The Ideal Orator and Literary Critic as Technical Communicators: An Emerging Revolution in English Departments. *Essays on Classical Rhetoric and Modern Discourse*, ed. R. J. Connors, L. S. Ede, and A. A. Lunsford. Carbondale and Edwardsville IL: Southern Illinois University Press.

White, J. 1985. *Heracles' Bow: Essays on the Rhetoric and Poetics of the Law.* Madison WI: University of Wisconsin Press.

Wickliff, G. A. 1997. Assessing the Value of Client-Based Group Projects in an Introductory Technical Communication Course. *Journal of Business and Technical Communication* 11:170–92.

Williams, S., B. Heifferon, and K. B. Yancey. 2000. Reflective Instrumentalism as a Possible Guide for Revising a Master's Degree Reading List. CPTSC: Available online at http://www.cptsc.org/conferences/conference2000/Williams.html.

Williamson, W. J., and P. H. Sweany. 1999. Linking Communication and Software Design Courses for Professional Development in Computer Science. *Journal of Language and Learning across the Disciplines* 3:103–6.

Wilson, R. 2001. A Higher Bar for Earning Tenure. *Chronicle of Higher Education*, January 5, A12–14.

Wise, J. M. 1998. *Exploring Technology and Social Space.* Thousand Oaks CA: Sage.

Wojahn, P. 2001. Blurring Boundaries between Technical Communication and Engineering: Challenges of a Multidisciplinary, Client-Based Pedagogy. *Technical Communication Quarterly* 10 no. 2:129–48.

Youga, J. 1989. *The Elements of Audience Analysis.* New York: Macmillan.

Young, A., and T. Fulwiler, eds. 1986. *Writing across the Disciplines: Research into Practice.* Upper Montclair NJ: Boynton/Cook.

Young, J. R. 2000. David Noble's Battle to Defend the "Sacred Space" of the Classroom. *Chronicle of Higher Education*, March 31, A47–49.

Young, R., A. Becker, and K. Pike. 1970. *Rhetoric: Discovery and Change.* New York: Harcourt Brace Jovanovich.

Zimmerman, D. E., and M. Long. 1993. Exploring the Technical Communicator's Roles: Implications for Program Design. *Technical Communication Quarterly* 2: 301–17.

Zuboff, S. 1988. *In the Age of the Smart Machine: The Future of Work and Power.* New York: Basic Books.

CONTRIBUTORS

TRACY BRIDGEFORD is Assistant Professor in the Department of English at the University of Nebraska at Omaha. Bridgeford coordinates UNO's Graduate Certificate in Technical Communication and teaches courses in technical communication, digital literacies, information design, and editing.

RICHARD (DICKIE) SELFE, Director of Computer-based Instruction in the Humanities department at Michigan Technological University, directs a technical communication-oriented computer facility supporting one of the largest undergraduate and graduate technical communication programs in the country. He teaches a range of computer-intensive composition, technical communication, and graduate courses.

KARLA KITALONG is Assistant Professor of Technical Communication at the University of Central Florida in Orlando. Her research interests include visual communication, technology studies, and usability. She teaches courses in the Honors College, the undergraduate and masters level technical communication programs, and the Texts and Technology doctoral program.

CHRISTINE ABBOTT is Professor Emerita of English at Northern Illinois University, where she developed the undergraduate and graduate programs in technical communication and the Institute for Professional Development, a partnership between NIU and the Chicago Chapter STC. She is author of *Technical Writing in a Corporate Culture: A Study of the Nature of Information.*

BARRY BATORSKY has taught technical writing for twenty years at NYIT, Polytechnic University, Tufts, and now at DeVry University where he also teaches dramatic literature and 20th Century Fine Arts. His Ph.D. is in Comparative Literature from the Graduate Center of CUNY with a dissertation on the theater of Brecht and Beckett.

GARY BAYS teaches composition and technical communication courses at The University of Akron Wayne College and oversees a certificate program in workplace communication. Bays currently serves as editor of *The Journal of the Ohio Association of Two-Year Colleges* and has worked as a freelance corporate writer for many years.

MARK CHARNEY, BETH DANIELL, THARON HOWARD, SUSAN HILLIGOSS, MARTIN JACOBI, CARL LOVITT, AND ART YOUNG participated in the founding of Clemson's MA program in Professional Communication (MAPC); they were joined later by CHRISTINE BOESE, BARBARA HEIFFERON, BERNADETTE LONGO, SEAN WILLIAMS, and KATHLEEN BLAKE YANCEY. Since the completion of this chapter, the composition of MAPC faculty has changed, with some colleagues leaving—Christine Boese to CNN, Beth Daniell to Alabama, Berndette Longo to Minnesota, and Carl Lovitt at Penn State Berks-Lehigh Valley—while others have joined us—Teddi Fishman, Morgan Gresham, Deb Morton, Michael Neal, and Summer Smith Taylor. Among our current projects is development of a Ph.D. in Professional Communication that we hope to begin offering in fall 2005.

STANLEY DICKS is associate professor in Technical Communication at North Carolina State University, where he teaches technical communication management, document design, and usability. His book, *Management Principles and Practices for Technical Communicators,* has been published in the Allyn & Bacon Series in Technical Communication.

JAMES DUBINSKY directs the Professional Writing Program at Virginia Tech. His teaching and research interests include service-learning and civic engagement, rhetorical theory, and the history of writing pedagogy. In 2000, he was awarded Virginia Tech's Student Leadership Award as the Outstanding Service-Learning Educator.

ELAINE FREDERICKSEN is Associate Professor of English at the University of Texas at El Paso where she directs the Bilingual Professional Writing Certificate Program. She earned a Ph.D. in Rhetoric and Composition at the University of Alabama. Her book *A New World of Writers: Teaching Writing in a Diverse Society* was published in January 2003.

JEFF GRABILL is Associate Professor in the Department of Writing, Rhetoric, and American Cultures at Michigan State University. He directs the professional writing program and co-directs the WIDE Research Center (Writing, Information, Design in E-Space). His current work focuses on how people use information technologies for research and participation in public decision making processes.

ANNMARIE GUZY, Assistant Professor of English at the University of South Alabama, teaches composition, technical writing, and gender studies. Her publications include "Returning Students and the Technical Writing Course" in *Kairos* (6.2 Fall 2001) and a National

Collegiate Honors Council monograph, *Honors Composition: Historical Perspectives and Contemporary Practices* (2003).

Craig Hansen is Professor and Director of the technical communication B.A. and M.S. programs at Metropolitan State University in St. Paul, Minnesota. He has published widely in the areas of technical communication, business communication, and composition. His professional experience includes technical and managerial positions in the computer industry.

James Kalmbach's teaching and research focus on technological literacy, computers and composition, web authoring, and technical communication. He has published in *Computers and Composition, Kairos, Journal of Reading, Technical Communication, Technical Communication Quarterly,* and is the author of *The Computer and The Page: Publishing, Technology and the Classroom* (Ablex, 1997).

Brad Mehlenbacher is Associate Professor in Training & Development (ACCE) and an Adjunct Faculty member in Ergonomics (PSYCH) at NC State University, and received his PhD in Rhetoric and Document Design from Carnegie Mellon University. He is co-author of the (1993) Ablex book, *Online Help: Design and Evaluation.*

Sam Racine earned her PhD in rhetoric and technical and scientific communication from the University of Minnesota. She is currently the Senior Usability Specialist at Unisys Corporation.

Laura Renick-Butera teaches technical writing at Nokomis Regional High in Newport, Maine. Previously, she taught technical writing and composition at DeVry Technical Institute in New Jersey. She holds a Ph.D. in English from the University of Pennsylvania.

Laura Sullivan teaches English and journalism, both print and electronic, at Granite City High School. Ms. Sullivan, using her working knowledge of technology, opens up the English classroom to communications, helping her students to better apply the curriculum to their lives and the world around them.

Phillip Sweany, Associate Professor of Computer Science at University of North Texas, has taught several courses where students work in teams to complete a significant software project. Writing submitted in such courses has led to a profound appreciation of the need for good communication skills.

Billie Wahlstrom is Vice Provost of Distributed Education and Instructional Technology for the University of Minnesota. Wahlstrom is also a professor of Rhetoric and teaches information design in the classroom and online. Recent publications deal with technology-enhanced learning and technology's impact on higher education, and she has published in many journals including *Computers & Composition* and *Quarterly Journal of Speech.*

BILL WILLIAMSON is Assistant Professor of English at the University of Northern Iowa, where he serves as the Coordinator of Professional Writing.

MICHAEL ZERBE is Assistant Professor of English and Humanities at York College of Pennsylvania. He teaches courses in first-year writing, writing in professional cultures, advanced composition, medical writing, editing, web design, and the history of rhetoric. He earned a Ph.D. from Purdue University, MTSC from Miami University, and BS (in chemistry) from James Madison University.

INDEX